THE BATTLE FOR BARRELS

"This book effectively undermines the validity of the theory of Peak Oil and comprehensively demolishes the arguments of its proponents. It is a 'must read' antidote to the gloom and doom conclusions of oil scarcity."
Peter R. Odell, Professor Emeritus of International Energy Studies, Erasmus University, Rotterdam (author, *Why Carbon Fuels Will Dominate the 21st Century's Global Energy Economy*)

"This is a serious work that serious readers should read, and Duncan Clarke offers a smart and insightful survey of one of the most intriguing issues of our times – Is the world running out of oil? – explaining why the doomsayers are wrong. He diagnoses the psychological mindset, the historical mistakes and the current complexities concerning the evaluation of how much oil lies beneath, and shows a positive view of our energy future."
Leonardo Maugeri, senior vice-president, Eni SpA (author, *The Age of Oil: The Mythology, History, and Future of the World's Most Controversial Resource*)

"Duncan Clarke provides a brilliant insight into the world upstream, Peak Oil, international corporate strategies, geopolitics, business, economics and technology. His unique worldwide experience provides an astute analysis of critical issues that clarify and interpret the historical future of oil and the modern world in the 21st century."
Dr Alfred J. Boulos, former senior director, International, Conoco Inc; former president, Association of International Petroleum Negotiators and European Petroleum Negotiators Group

"This book portrays an independent mind and Duncan Clarke's analysis and interesting opinions shed much light. Few things are trickier than reserve estimates. We had the rule: 'You will never know the reserves of a field, before it is depleted ... and even then.' What about reserves of not yet discovered fields? Gerry Dixon, former president of Petroconsultants, always said: 'If we badly need it, we will find it.' Clarke's optimism on the world is justified. One day oil may be depleted, unless humanity does not need it before."
Christian Suter, CEO, Petroconsultants, 1987–98, and member of the board 1968–99

"*The Battle for Barrels* lucidly shows that not all is known on potential oil reserves-in-ground, and Peak Oil is very wrong on its view that such reserves are now fully known and finite. Clarke credibly demonstrates that there is much conventional oil to be discovered while economics and above-ground

factors such as global access and resource nationalism, plus technologies, will shape real oil world futures in ways unlike Peak Oil theorists presume, especially given the Earth's vast oil endowment."
Fred Dekker, managing director, Wessex Exploration; former vice-president, Asia Pacific New Ventures, Unocal Corporation

"Peak Oil has caught global attention because the media love catastrophic news. Duncan Clarke, who has had close contact with top leaders in the oil industry and its best experts over three decades, has demystified this dogma and shown the crude realities. Clarke demonstrates convincingly that the world's recoverable reserves, eventually limited but more extensive than generally perceived, will be shaped not only by geology but by future strategies, long-term oil prices, new technologies, dynamic markets and shifting geopolitical access."
André Coajou, former senior executive, Elf Aquitaine (Exploration, New Ventures, Negotiations)

"*The Battle for Barrels* is a first-class treatise on the myriad ideas and 'almost theories' on the end of oil and its impact on our society. It separates facts from sensationalism, analyses all relevant ideas, provides rationale to our understanding, and makes a brilliant contribution to the field of energy economics."
Professor Edmar de Almeida, Instituto de Economia, Universidade Federal do Rio de Janeiro

THE BATTLE FOR BARRELS

Peak Oil Myths & World Oil Futures

Duncan Clarke

P

PROFILE BOOKS

First published in Great Britain in 2007 by
Profile Books Ltd
3A Exmouth House
Pine Street
London EC1R 0JH
www.profilebooks.com

Typeset in Times by MacGuru Ltd
info@macguru.org.uk
Printed and bound in Great Britain by
Clays, Bungay, Suffolk

A CIP catalogue record for this book is available from the British Library.

ISBN-10: 1 84668 012 3
ISBN-13: 978 1 84668 012 0

The paper this book is printed on is certified by the © 1996 Forest Stewardship Council A.C. (FSC). It is ancient-forest friendly. The printer holds FSC chain of custody SGS-COC-2061

FSC
Mixed Sources
Product group from well-managed
forests and other controlled sources

Cert no. SGS-COC-2061
www.fsc.org
© 1996 Forest Stewardship Council

Contents

For Babette

Acknowledgements

There is a cast of many I must thank for their direct and indirect contributions over the years to the ideas and efforts that have borne fruit in *The Battle for Barrels*. It is not feasible to cite all of them here, but for all those not mentioned an apology is in order. All should remain blameless for any errors of commission or omission in the text.

Some early ideas on non-linear concepts were drawn from scientists at the Centre Européen pour la Recherche Nucléaire (CERN) in a series of casual discussions in Geneva in the mid-1980s, accompanied by some *vin ordinaire*. They led me to rethink the theory and practice of conventional economics. My initial application of notions of chaos and complexity under uncertainty to the world upstream is to be found in *Strategic Petroleum Insights*, published by Petroconsultants in the late 1980s – a work that sought to tackle the notions of reserve growth and world oil endowment in dynamic terms. This thinking was enhanced by an updated Global Pacific & Partners report published under the same title in the early 1990s. Non-linear ideas have been a template for most of our concepts, research and advisory work since that time.

Christian Suter (then CEO of Petroconsultants) along with the former president, Gerry Dixon, encouraged and supported many of the original ventures undertaken for and with that firm. The relationship had been built on an association with the irrepress-ible Harry Wassall from the early 1980s until this legend of the oil industry passed away. Both before and during 1985–89, when I was with Petroconsultants, many colleagues in the Economics Division and other parts of the firm played an important part in my acquisition of knowledge about the economics and complex-ities of the international upstream industry.

By this stage I had also been influenced by a wide range of writers, thinkers and other individuals – many of them from discip-lines or professions outside economics and the oil industry – with whom I had good fortune to cross paths. Michael Holman, Africa editor of the *Financial Times* from 1984 to 2002, has remained an inspiration to us all. The late David Mitchell taught me much on management and the art of strategy. And there have been many from the world of economics who have shaped my notions of applied economics. Here I would like to thank in particular Rob J. Davies for many a long discussion on diverse themes over the decades of our friendship. Heidi Holland introduced me to the Okavango, where an outstanding guide, "Jules of the Okavango", managed to reinforce and "make me see" the realities embodied in the theories of the natural non-linear.

Inside the global upstream industry, my indebtedness extends to so many people all over the world that a separate work would be required simply to cite and thank them. However, I would like to acknowledge my appreciation of some technical and geoscience colleagues known for many years – in particular, Fred Dekker (ex-Unocal), André Coajou (ex-Elf and Total) and Bill Brumbaugh (ex-Mobil), and Andrew Lodge and David Slocombe

(Hess) – for sharing their varied and considerable accumulated wisdom drawn from hands-on experience worldwide. This debt of gratitude likewise extends towards many other individuals in the international exploration business from a multiplicity of corporate and government entities.

It was a chance encounter that led to Profile Books. Many thanks are due to Andrew Franklin (publisher), Daniel Crewe (editor) and their skilled team for their commitment to this book. A special appreciation is due to Anthony Haynes, an exceptionally gifted editor, for insightful, wise and knowledgeable treatment on edits to much improve the original draft, style and presentation.

Lastly, heartfelt thanks go to our excellent team at Global Pacific & Partners, including Tim Zoba Jr, our former senior partner (1989–2003), all of whom helped shape our worldwide oil industry knowledge and relationships. They all deserve much recognition for their considerable skill and efforts made over so many years – especially Babette van Gessel (group managing director), who continues on this journey with me after outstanding contributions to our firm and to so many players in the world upstream over the past 14 years.

Duncan Clarke
October 2006

Introduction

This is a treatise on the movement and theory known as Peak Oil. It examines their predictions concerning the so-called "end of oil" – even the "end of civilisation" – and the fashionable global angst related to them. It draws on the author's involvement in the international exploration business since the late 1970s – experience that has been shaped by exposure over several decades to economics and by an extensive advisory practice in upstream business strategy ("upstream" being the part of the oil business devoted to exploration and production).

The diagnosis is grounded on ideas of non-linear dynamics and complexity. It illustrates the role of known and unknown elements in the history and future of world oil to illustrate that Peak Oil's doom and gloom scenarios are mistaken. The treatise demonstrates that, in the light of complex global upstream realities, we can be much more optimistic about the future of world oil than the Peak Oil movement would have us believe.

In the publications, media comment and internet discourse that comprise public debate, it is widely accepted that worldwide discoveries of conventional oil have peaked, that oil production either has declined or will soon do so, and that the era of cheap oil – or even of abundant oil – has gone for ever.[1] There are, of course, different views within the Peak Oil movement, especially on the dates contained in their forecasts. It is the core idea – what below is termed the Peak Oil model – that is addressed here. The aim is to provide some seasoned, analytical, heretical and polemical reflections on Peak Oil's debates and on its architects, adherents, allies, activists and critics.

A key aim of this book is to raise questions about the fundamental concepts of Peak Oil theory and the presumed truths contained in its increasingly accepted conventional wisdom. These questions shed new light on the answers implicit in the Peak Oil model. This approach raises the possibility that new questions may prove more important than existing answers. This inquiry examines the issues and data at the core of the Peak Oil model and its claims of supposedly diminishing world oil patrimony accompanied by an aggravated view of near-term oil and energy crises. These claims are based on restrictive and pessimistic geotechnical paradigms that, in effect, entail the wholesale rejection of the discipline of economics and its findings. This study takes a hard and irreverent look at the often heavily dramatised post-peak socioeconomic, political and human crises forecast with unbridled certainty for the early 21st century.

The treatise is intended to be neither academic nor definitive. The ideas (both central and derivative) and the social phenomenon associated with Peak Oil remain to be discovered more fully. This book does not attempt a statistical refutation of the Peak Oil model. Nor does it provide a competing model estimating the supply future. Rather, it offers a commentary on both the prolific literature now found on Peak Oil and the parties most associated with it. It is an examination of Peak Oil's theory and evidence. I ask whether hidden or unseen assumptions exist in the Peak Oil model. In exploring Peak Oil from both inside and outside, this book takes the first steps along the road towards a fuller understanding of the theory and its social ecology.

Understanding the full implications of the Peak Oil thesis requires perceptions from multiple angles. It is a holistic view that we require – one that reviews not only the theory and the supposed evidence for it, but also the anthropology of these ideas, their genesis and the wider contexts of the world oil debate. This approach will help explain the global angst that Peak Oil produces, the passions it evokes, and the meanings assigned by its constituents. Peak Oil – as idea, event and process – has an intellectual and a social context. Its concepts and the social dramas around them should to be seen together: they need, so to speak, to be placed in a worldwide cultural matrix.

Most importantly, the Peak Oil model provides specific technical reasons for the supposed depletion of the world oil supply in the near future. These derive primarily from a certain geological determinism, according to which oil "reserves" (a term in need of some definition) are uniquely fixed and finite by the limits of nature. On this view, remaining oil reserves are now smaller than once thought and depleting fast, creating an irreversible and fundamental discontinuity for future supply. It is as if there is some "skeleton in the oil kitchen". Many, if not all, potential offsetting or mitigating conditions (solutions providing greater reserve access, for example) are perceived to make little, if any, difference to the catastrophe ahead. The crisis will occur, we are informed, come what may – regardless of markets and economics, rising crude oil prices, new or future technologies, potential improved exploration acreage and/or access to restricted world oil zones, changes in government policies on resources, new corporate strategies, development in unconventional oils, growth in diverse renewable fuels, or the like.

According to Peak Oil, the "oil world" that we now inhabit is doomed. This "end of oil" moment, coinciding roughly with the close of the 20th century, heralds the termination of a grand epoch – even the "end of civilisation" as we now know it. Our rapid descent into the looming, unprecedented PetroApocalypse is preordained, a Nostradamus-type event. It is written in the stones, if you like. For the dedicated followers of Peak Oil, there will be no nirvana, or even a return to our recent, oil-addicted, past. Before us lies a world of political and economic hyper-collapse, endemic wars targeted on scarce oil resources, and dramatic, worldwide impoverishment. Many have claimed that this depicts the landscapes of the oil world even now.

How has this situation arisen? According to Peak Oil theory, it is because our contemporary world is caught between accentu-ated oil addiction on the one hand, and the constraints of supposed "geological controls" and preordained limits to crude reserves on the other. Moreover, our public and private reserve estimates and databases seriously inflate (some say malevolently) the volume of oil that remains available to us. Global market users have become dangerously dependent on unsustainable growth in oil demand at a time when supply from the world's *quantum fixe* of oil is in a state of seriously aggravated depletion.[2] In a nutshell, the world is up the creek without the necessary paddle.

According to Peak Oil thinkers, oil supply will rapidly decline towards eventual exhaustion. The world will be in hock, politically and economically, to a few rather questionable Middle Eastern oil suppliers (who may not even have the oil reserves they claim). The price of crude will rise radically, inevitably and permanently. We will be in crisis, suffering from the curse of decades of humanity's mismanagement of oil usage and God-given oil depletion patterns. Hence a dangerous and devastating future awaits us all, soon. But eventually, following a dramatic and traumatic transition, a "New Age" will arrive, whereupon from the ashes of the coming oil crisis will emerge a sustainable energy world and planet, as mankind lives out the diminished life span left in the fossil fuel record.

Is this our predicament? Will the future really be like this? What if the paradigm and the conceptual underpinnings, the model and its data, are wrong?

Parts 1–5 of this treatise present the central notions of Peak Oil, the relevant criticisms and a set of contrary ideas on world oil realpolitik drawn from oft-forgotten or much-dismissed insights from technical and non-geological quarters. These ideas provide a wider context to the debate on the alleged oil crisis and, as it turns out, belie the supposed immediacy and probability of Peak Oil's doomsday dramas. In the Peak Oil model we find several evolving scenarios, shifting sets of supposedly "absolutist" data and speculative futuristic images, some tilting towards the extremities of absurdity.

There is a voluminous literature on Peak Oil issues from a wide variety of technical, academic, institutional, NGO, private and professional sources. They display multiple pedigrees and angles of approach: geoscientific, technical, statistical, economic, social, political, activist and even quasi-philosophical. This treatise does not survey all sources or deal with all the questions raised as if it were a definitive account. It focuses on core claims and critical essentials, with comments on those issues of wide concern to society and the global oil industry.

The key notions here are the phenomenon of Peak Oil itself, the parametric models that it uses and their formulaic methods, the internal and external flaws, the social drama and the meanings that it produces concerning the world's future, as well as the claims, rationale and implications of its proponents. Peak Oil thinking has become a moving target. Over the years the many changes in predictions and data have clouded the supposed precision of Peak Oil forecasts. The theory presents a body of ideas with inner complexity. The assumptions buried deeply in several concepts need to be unravelled. Many questions arise about the model's assumed veracity and its presumed implications for society.

Part 1 recalls the wide and elastic boundaries in human adaptation and thought. It draws attention briefly to the origins of man and the skills in tracking and the accumulated knowledge that enabled the long-term survival of the Khoi-San in the Kalahari. This starting point may seem obscure. But imagination, along with man's adaptation to nature, has long played a crucial role in humanity's survival. Stone Age economies possessed a power of adaptation that modern societies might emulate and bring to bear on their own survival, if the end of oil and PetroApocalypse (a metaphor for serial large-scale oil crises) were now at hand.

Part 2 deals with the social dramas associated with the end

of oil and of civilisation and the PetroApocalypse. It focuses upon the collage of thinkers on whose ideas these images are built, and their views on the future. It sheds light on Peak Oil's latest thinking, gleaned from its 2006 annual conference in Pisa, attended by the author. That conference dwelt on the coming oil and social dramas, an abiding theme in Peak Oil theory. Several end-of-oil syndromes amplify the warnings frequently issued, the global angst generated and the nature of our impending doom. Many individuals and groups inhabit this social space and there are various cults of concern among theorists, advocates, activists and politicians of different types.

The debate has moved into the political world and the media, where it has received much exposure. Perhaps it has even generated a herd mentality. Many people now wish to shape the next world oil/energy order, in ways very different from those the world is accustomed to. Will this new dogma concerning the social drama associated with Peak Oil be confounded by societal adaptation, the power of economic paradigm shift, or geological and managerial imagination – or will world oil reserves remain static and deplete inexorably, so that a meta-crisis will envelope all?

Part 3 focuses on the roots of the Peak Oil social drama: the knowledge and mindsets that provide the foundations on which Peak Oil theory is built. It asks whether linear thinking is suited to diagnosis of the issues, let alone prediction far into the future. Why does Peak Oil hold a narrow parametric set of views? Our most advanced ideas on economics and in science have rediscovered the critical importance of non-linear ideas for interpretative understanding. This set of tools is essential for discerning the future of world oil and even measuring the status quo in reserves and supply. Oil is not much different from any other resource, though this is much disputed by Peak Oil thinkers. Many lessons of history can and probably will be applied to managing its future evolution. It is essential to understand the fundamental model, concepts, technical claims, assertions and data selections of Peak Oil on its own terms. Its framework needs to be tested both inside and outside the singular model of oil depletion, and its supposedly indisputable numbers placed in wider context. There are Peak Oil gurus who deserve special attention, with their prolific writings and forecasts on Peak Oil, based on ever-shifting dates. A rich cast of characters is encountered guarding the gates of Peak Oil.

Part 4 examines the contemporary world's upstream realities. It highlights several deep-level conditions that contribute to the global oil game today and clarifies the reasons behind the current state of affairs. Global estimates of oil reserves and future supply are not determined by any one factor alone and so a linear view is unsuitable for understanding our current dilemmas. Governments are central to the problems of access to oil reserves for development and production and to the acreage supply on which future exploration depends. A multitude of companies (both large and small) also play a critical role. Many independent players form a sort of long corporate tail. They provide a cutting edge for the world exploration future. The strategies of corporate oil (a term used for the entire industry in private hands) reflect rational adjustments to new realities and a world of increasing risks, threats and uncertainties.

We need to test the baseline oil numbers on which Peak Oil grounds its apocalyptic claims. Though its pessimism about the future of global exploration is strong, much is unknown about exploration potential – especially in Africa, Asia, Latin America, the Middle East and Russia and frontier zones across the world. This is where most future oil reserves exploitation and undiscovered oil is to be expected. We also need to consider potential reserve growth and future technology development. Many Peak Oil adherents dismiss economic considerations categorically. For them, every-thing is determined below the surface (where, it is claimed, all is known): the above-ground (of which our knowledge is certainly not absolute) is held to be irrelevant. Management and technology are considered merely neutral in the future oil equation.

The oil world has never been immutable. What is known today is likely to be no more than a small fraction of what might be known tomorrow. Yet Peak Oil claims an exceptional status for its restricted geological knowledge. It even claims that oil is "not just a commodity". These claims clash, in intellectual and policy terms, with traditional economic theory (especially with regard to the role of dynamics). Many Peak Oil ideas are likewise at variance with managerial and geoscientific practice within corporate oil. Companies have moved more deeply into complex, developing oil worlds for reasons of opportunity and strategy. This is a world of threats and risks with sometimes restricted reserve access, exploration, discovery opportunity and production. Many new strategies are in play in this arena. Is this evidence of an imminent depletion crisis – or does it reflect new geopolitical maps that are reshaping the global oil industry above the ground?

In the 21st century there are many hydrocarbon games in play, with "new players" involved in the global scramble for oil resources. The great oil game in process now and for the foreseeable future creates uncertainties and angst related to the perceived and actual complexities in the worldwide search for oil. Resource nationalism is the latest problem to have an impact on exploration and development. Peak Oil interprets all this as evidence of an imminent peak in oil production, but is this correct or is there a better explanation?

Part 5 considers the thoughts of Machiavelli, esteemed for his thinking on strategy and an arbitrator of complex and labyrinthine worlds. It shows that Peak Oil advocates have intuited the future not through the wide-open doors of lateral perception but via the restrictive lens of a single microscope. Peak Oil discourse reflects its narrow choice of theory and models, as much as the pure numbers and the conclusions to be drawn from them. Does the essential crisis of Peak Oil merely reflect a condition like some twilight in the mind?

The text of this treatise comprises a series of vignettes. Endnotes identify key sources and provide further observations. Though Peak Oil texts tend to be replete with tables, graphs and formulae, I have included none here. So many images and projections have been deployed that trawling through them all would tax anyone. I have, however, made a genuine attempt to represent arguments fairly and, where possible, to indicate the kinds of evidence that they draw on.

There is no special methodology employed. I have read a multitude of texts written on this issue over many years, interpreted them to the best of my ability, reviewed the critics and the positions they adopt, and listened closely to Peak Oil debates and its interlocutors. Several texts and conference presentations have proved instructive, notably those generated by Peak Oil thinkers themselves, as have insights gleaned from several world-knowledgeable colleagues in management and geoscience. I have drawn also on my own experience working in the world upstream industry and practising economics (both inside and outside the industry) with governments, national oil companies and oil companies in the developing world and elsewhere.

The endnotes point to many of the key texts and authors in this flourishing worldwide debate, to the principal institutions, both official and unofficial – notably the Association for the Study of Peak Oil (ASPO) and several state and international bodies of recognised status in world oil – as well as a range of published and unpublished works (many found on the internet) that cover the span of arguments in this debate. The Notes on further reading provide additional guidance. Appendix 1 contains a list of abbreviations and measures, and Appendix 2 illustrates the conception of resources/reserves used by the much-respected Society of Petroleum Engineers (SPE).

Part 1

THINKING UNTHINKABLES

There are those for whom imminent Peak Oil is now thinkable. Yet for others it is still unthinkable. For some the thinkable is restricted to the limited or finite view of resources and oil reserves. For others the near-unlimited resource boundary is even thinkable. And there are those for whom the subtle combine of the thinkable and as yet unthinkable may shape the course of world oil history.

So at the outset it is worth asking questions of Peak Oil, its theory and evidence as well as its historical traditions and milieu. These questions should be set inside a wide canvas of global history. Here there are analogues of human experience on which we can draw even if they are not the obvious ones. One lesson they provide is that of societal adaptability under conditions of adversity.

The knowledge held in the world, including the oil world, is also vast even within a single discipline or intellectual tradition, and so it is best to operate with the widest optics. This is essential to the task of understanding the world oil future, since an enormous range of geopolitical and economic forces conspire to shape its present and will do so in its future.

Yet Peak Oil holds that our future is determined and that we know all now that it is necessary to know about why it will be so. Many others believe the contrary. For them all is not known, let alone the subsurface fossil record on which much may even be unknowable for some time. In consequence we are left with sharply contesting sets of views about both global society and the world oil future.

In essence, one view holds scant regard for human adaptability even within apparent oil crisis and the other envisages a world of continuous adjustment in both society and the oil industry that will suffice to surmount the challenges ahead. Which view will prevail?

1

Questions on Peak Oil

If the moment of Peak Oil is either already with us or just around the corner, then radical change is afoot. Soon our current ways and means will no longer suffice. Man, some commentators have suggested, is at a unique turning point, a great "unfolding" of a kind that we have never confronted before and are wholly unprepared for. Through a series of crises we will descend, rapidly and turbulently, from our current peaks of material accomplishment. Others contest the prophecies of doom and gloom. Within these debates are clashes of ideas, disciplines and schools of thought. Though these prophecies are supposed to derive incontrovertibly from geoscience, many geoscientists working for oil companies dispute them.

If the geoscientific prophecies of doom are indeed unchallengeable, then our 250-year-old tradition of economic thought would seem to be redundant. Peak Oil, according to one of its gurus, "probably heralds the End of Economics as currently understood".[3] Adam Smith, David Ricardo, Alfred Marshall, John Maynard Keynes, the schools of classical and neoclassical economics – all will be consigned to the dustbin of history.

Unsurprisingly, economists and investors, especially those involved in the upstream business, tend to take a more sceptical view.

Questions regarding the implications of Peak Oil may well be long-term ones. Here social science is notoriously weak. The Peak Oil model, however, is extraordinarily ambitious. It derives from the projection of a single variable – the supply of oil flowing from narrowly estimated reserves – and, moreover, reduces this to a determinant purely of geology. Winston Churchill may have said that the chain of destiny can only be grasped one link at a time, but the Peak Oil model exhibits no such circumspection. As over the next decade, according to the model, oil supplies become rapidly depleted, so the world starts to hurtle down a precipitous slope into disorder.

We need a sense of curiosity and sufficient initial ambivalence to probe the model's strengths and weaknesses. Questions both broad and specific need to be explored. Yet many Peak Oil adherents assert that they are correct in all essentials. How, we may ask, can they be sure?

Perhaps, however, there is doubt in some minds. After all, the predictive capacity of the model has proved uncertain: the dates forecast for various events have needed to be revised. The model may, therefore, have been poorly designed. It depends largely on some selective statistics, while the real world is shaped by history. The social visions produced by Peak Oil also show little regard for historical evidence, deriving more from a heightened sense of the dramatic.

Given the inadequacy of their quantitative forecasts, we might well ask why Peak Oil adherents have so much difficulty letting go of them. Their exclusion of all above-the-ground factors (economics, technology, strategy, policy) makes them seem even more dramatic. Perhaps its psychological appeal derives from its abstraction from the complex world. Maybe Peak Oil's advocates have simply invested too much pride or intellectual commitment in the model to be able to abandon it. Perhaps too the model has been nailed too securely to some technical or political masts.

To an external observer what is remarkable is the obsessive nature of some Peak Oil advocacy. In it there is found profound inflexibility, a marked reluctance to adapt to external influence, as if its adherents find the model captivating and all-embracing. Certainly they fail to consider data sources that elsewhere are regarded as standard. It is as if all government institutions and leading corporations form one vast web of misinformation, delusion, idiocy and incompetence. The result is a self-separation of Peak Oil from the mainstream oil industry. It ends up preaching mainly to the converted.

We are left with two sharply contrasting sets of ideas. On the one hand, there is deep pessimism over mankind's ability to adapt, even within an oil crisis. On the other, there is the historical evidence of human change and adaptation, both in socioeconomic terms and in the dynamics of the upstream industry.

2

Kalahari dawn and the art of tracking

In the ancient hunter-gatherer economies, our forebears functioned with very low levels of energy per head per year. We are apt to look down on these societies. We tend to lump them into a "black box" labelled "subsistence economies". Our image of them is a simple one of Palaeolithic poverty. Yet it has been suggested by Marshall Sahlins that we should see these peoples as the world's first affluent societies.[4] Despite extremely limited resources, such societies managed to strike a balance with nature that enabled them to sustain themselves.

While we may marvel at the ability of primitive societies to survive, many are haunted by a fear of returning to Palaeolithic barbarism. Such fears lie behind much of our angst about the supposed prospect of a future without oil. They tempt us to look for radical means to salvation. The spectre of a return to the Stone Age fuels the clamour over the message of Peak Oil. According to Peak Oil purists, nothing above the ground can fix the situation. Our future is geologically determined: the New Stone Age draws inexorably closer. What, in the circumstances, may be learnt from the societies of the original Stone Age?

Probably the world's original human inhabitants, the Khoi-San (more widely known as the Bushmen) of the Kalahari lived for millennia in an environment of extreme resource constraint and minimal energy levels. They survived by endlessly adapting to their environment.

The Khoi-San's creativity consisted not purely in external discovery but in their intellectual curiosity. Crucially, they developed the art of tracking. Tracking enabled them to discern and locate critical scarce resources (in this case animals) even when they were unable to see them. Louis Liebenberg has suggested that in the development of tracking may be seen the true origin of science.[5]

The development of the Khoi-San's thinking remains relevant today when, in place of animals, energy (especially oil) has become our critical scarce resource. In our search for energy, the example of the Khoi-San warns us against presumptive boundaries and encourages us instead to apply to our apparently finite world the elasticity of human ingenuity.

Let us consider a little more closely what was involved in the development of tracking. Tracking required the unrelenting and systematic gathering and interpretation of data. It required deep thought based on a comprehensive appreciation of both measurable and non-measurable elements in the environment. The Khoi-San had to learn to read marks and footprints – to discover in them hidden insights into the movement of animals.

To do so required knowledge of the environment in general

and also of the strategic and tactical behaviour of herds and of particular animals. It required too an understanding of the role that each animal played in the environment and the complex inter-relationships involved. It necessitated too a sense of wide-lens perception and dynamic adaptation – a willingness to recognise changing scenarios and to adjust to them. The Khoi-San needed to appreciate the non-linearity of nature and to develop their own chaos theory to survive.[6]

The development of tracking required not just knowledge, but also – crucially – imagination. The Khoi-San had to speculate on the movement of animals and to generate hypotheses on such matters. Trackers needed, as it were, to become the animals that they were tracking – to learn to see the environment through the eyes of their prey. They were forced too, for their own survival, to learn to rely on their presentiments (or pre-sentiments) and to develop a sense of risk. Tracking was not merely a matter of spoor recognition or registering marks on the ground, but also of applied ingenuity and developing a refined intuition. This led to deep-level knowledge beyond mere instinct.

What then – to make the allegory explicit – could we learn from the Khoi-San about our approach to another world, which would be that of the upstream oil industry? Superficially, the two worlds are very different. Looked at more closely, however, the similarities are striking. To infer the future of oil, we too need not only to acquire empirical knowledge, but also to apply our imagination. Our future is not written in the rocks any more simply than that of the Khoi-San was written in the arid Kalahari sands.

For this intuitive appreciation, more is required than mere

measurement, technical knowledge, or linear extrapolation. We too need to look at the world with the widest possible lens – one that will enable us to see the complexity of dynamics and inter-relationships, both within the industry and world markets and in its relationship with the changing environment. We need to develop our curiosity and learn to adapt to new discoveries and to the turbulence in our natural and social, economic and industrial worlds. The example of the Khoi-San suggests, above all, that we need to recover an appreciation of the non-linear world.

From the example of the art of tracking, then, we can see that the ultimate boundaries in oil and resource-use adaptation may be our own. Our world is not simply limited by present measure: it can and needs to be creatively constructed. All considerations need to be found a place within an overarching interconnectivity. A purely quantitative model, presuming that all of importance is already known, has serious limitations. It was the examination of, and judgements about, the unknown that completed the Khoi-San trackers' knowledge and enabled their societies to function and survive.

In following the trail of the Peak Oil movement, Part 2 of this book applies these lessons. It asks: What is the habitat of this animal? In which environment does it find nourishment? How is Peak Oil related to the social context around it? What should we make of its prophecies for the world once it has reached its Peak Oil moment?

Part 2

THE "END OF OIL" AND
THE "END OF CIVILISATION"

Thanks to vivid media coverage and a prodigious output of publications, Peak Oil has begun to capture the public imagination. Within the oil industry, however, its theory is either dismissed or not so well appreciated. It has only rarely been subjected to rigorous analysis, although much evidence to contradict its theses is found. Its social model about the consequences of imminent peaking in oil has likewise been much neglected. Some take it as inevitable. It is as if history has been allowed to stop. In tracking the Peak Oil movement, we should attend to that theory and also its social ecology – its architects, organisation and milieu.

The Peak Oil movement combines genuine technical and scientific knowledge (at least on the part of some of its adherents) with a prophetic concern for the future of mankind. There is a missionary quality to the movement. It has, in particular, much zeal to testify to the death of our ailing civilisation. To understand this movement we need to consider the technical arguments and to evaluate its social and political dimensions.

Many of the ideas found within and around Peak Oil have the quality of undiluted pessimism. The media have amplified many of these dramatic elements. Reasoned arguments to the contrary are often then dismissed in public debate so that discussion becomes focused at the extremes.

It is, however, important to understand the sources of angst about Peak Oil and to examine the various arguments made by the adherents, including their dramatic predictions. This sheds light on the social and economic outcomes that flow from the technical models on which Peak Oil theory is built. Such models are typically sophisticated, mathematical and statistical. All this lends credence, in the eyes of the non-oil world, to their proclaimed results. Since the technical model is linked to the social model, each plays a part in the symbiotic relationship found. Both are a cause of much angst.

3

PetroApocalypse now

Some Peak Oil thinkers confine themselves to technical and scientific concerns – such matters as the timing of Peak Oil and the consequences for the derivative energy markets. Others focus explicitly on the social consequences. What unifies the Peak Oil message is the sense that it provides a grand awakening for a world that is seen as soporific and intensely vulnerable.

Though the outside world regards the Peak Oil movement as pessimistic, the movement sees itself rather differently: that is, as a group of alert and well-informed idealists or, even, as the only true realists in the debate. Within that position, however, there is a spectrum of views. While some see in the depletion of oil an opportunity for social transition and development, many prophesy an imminent PetroApocalypse. The doomsday quality of much of this thinking is clear from the titles and content of many popular books: *The End Of Oil*, *Half Gone*, *The Final Energy Crisis*, *The Long Emergency*, and so on.[7]

Peak Oil thinkers use a range of technical, quantitative and geological tools to construct a multitude of Peak Oil scenarios, all centred on fast-approaching oil depletion. Such forecasting has itself become a boom industry, growing as fast as it claims oil to be diminishing. We are warned not only of the depletion of oil reserves, but also of the consequences: plummeting living standards, social and political disorder, societal collapse (including, of course, the destruction of suburbia). There will, we learn, be an era of global oil wars – one that some say has already started. No one will escape from calamity. Out of the turbulence, however, a new, more harmonious world in balance with nature may eventually emerge.

Where, we might ask, are the origins of this curious movement? And how has it come to emerge from its roots into its current manifestations?

M. King Hubbert, who may perhaps be thought of as the grandfather of Peak Oil, was an American geoscientist with a long and distinguished career in research, including positions at Shell Oil and at the United States Geological Service (USGS; 1964–76). In 1956 he developed a method for estimating, some claim predicting, Peak Oil production in the US. "Hubbert's Peak" generated vigorous debate within the technical and policy worlds concerned with oil. Hubbert was ultimately to become a folk hero of the conservationist movement. In the 1990s his methods were adopted by the founders of the Peak Oil school.

Hubbert's methodology was far from clear. In *Hubbert's Peak*, Kenneth Deffeyes notes that Hubbert first reached his conclusions using guesstimates and then searched for data and methods to refine and support his conclusions.[8] Out of this messy method emerged an elegant forecast in the shape of a nicely symmetrical bell curve. This model may or may not have applied to production within the mature oil zone of the US. Nonetheless, many US players have obviously sought to explore and develop oil reserves abroad. There is, moreover, no reason to assume that the curve applies to all less mature oil zones outside the US.

Subsequently, a multitude of theorists have attempted to apply other types of "fitted curve" to global oil production. A mini-industry of statistical analysis has grown out of the application of Gaussian, Lorentzian, logistic and parabolic fractal curves to Peak Oil forecasting. Deffeyes himself estimated that, for global oil production, a Gaussian curve provided the best fit. He did, however, question how well the curve would serve for the purposes of prediction. He argued that, within the model, adjustment would need to be made to minimise internal inconsistencies. Beyond the model, guesstimates were required regarding reserve stock and production flows.

The results yielded by such models have proved sensitive to changes in data inputs. Views on the degree of fixity in oil reserves and on stock/flow relationships differ greatly. So, as a consequence, do the forecasts produced from them. Any one set of results should, therefore, be treated with caution. Yet such scepticism tends to be swept away by the certainty with which Peak Oil's governing assumptions are held. Central to the arguments of Deffeyes and his colleagues is a belief in the *quantum fixe*. Somehow, it seems, the volume of accessible discovered and undiscovered oil is absolutely finite and can be adjudged now and for ever. It is taken as a matter of geological fact that *quantum fixe* cannot ever alter and that little can be expected of future oil exploration.

If there is a contemporary guru of the Peak Oil lobby, it is Dr Colin Campbell.[9] He is a man who has gathered a cult-like following.[10] Indeed, he is no ordinary layman. A geologist with a PhD from Oxford University, he worked in the industry for nearly 40 years as a field geologist, manager and consultant. Since retiring in 1989, he has become the most influential thinker on Peak Oil, authoring several books and over 150 papers. The Oil Depletion Protocol, devised to guide the world through the PetroApocalypse, stems from Campbell's work. It is Campbell who is most widely quoted in the media as a spokesman on Peak Oil issues.

Campbell cut his teeth on technical matters while employed by Amoco, where he worked on the estimation of the world's oil resource. It was, it seems, while working for Amoco that Campbell's original pessimism set in: "I began to have a doomsday view of the future. We were, I came to realise, approaching an abyss".[11] In the late 1980s he worked on similar questions in an evaluation done by the Norwegian Petroleum Directorate. This was followed by a report on the world's oil resource, written with Jean H. Laherrère and published by Petroconsultants in 1995.[12] Campbell went on to found the Association for the Study of Peak Oil (ASPO), of which he remains honorary chairman. He is also a trustee of the Oil Depletion Analysis Centre (ODAC), a UK-registered educational charity.

The work that most fully reflects the scope and depth of Campbell's thinking is *Oil Crisis* (2005). This semi-autobiographical book contains much technical and quantitative data, wide-ranging comment and observations, and an extensive bibliography of published and unpublished texts. *Oil Crisis* gives us the essence of Campbell's case: time is short, midpoint depletion is upon us; official estimates of oil reserves lack credibility, social crisis is imminent. We are moving into the second half of the age of oil, a time of great international tension and one in which conventional economics will no longer apply. Provided the transition is properly managed, however, a better and more sustainable society may arise out of the crisis.

Such writings are permeated with a sense of earnest endeavour and mounting frustration. The world, it seems, is ignoring the scientific evidence: international and governmental institutions such as the International Energy Agency (IEA), the US Department of Energy, the USGS and the Energy Information Agency (EIA) and oil companies such as BP are in denial. The future scenarios that they publish are broadly wrong and hence probably worthless.

The assertions made in *Oil Crisis* emerge from a mixture of analysis and estimate. There are interviews with old oil hands including some original explorers, now long gone. The main foundation for the book, however, is Campbell's quantitative model and the estimates on which it is based. The finite oil stock is all-important. Technology has, in effect, reached the limit of its potential: any further application or development in technologies will merely accelerate production and hence depletion. The thinking of old flat-earth economists such as Maurice Adelman, Peter Odell and Michael Lynch (see Chapter 14), and economics in general, will be of no avail. The moment of Peak Oil is imminent, though the forecast for the year of the peak, once predicted to be some time back, seems now to vary from 2006 to 2010.

The unavoidable decline in oil production is only part of the doomsday scenario. Even a decline at a modest rate will undermine the foundations of our financial and industrial system. In a world of finite oil resources, we are trapped in a zero-sum game: one man's barrel is another man's loss, this extended to the idea that "one man's wealth has to be another's poverty".[13] Amid the tensions caused by inequality of wealth, the world system will collapse. Indeed, the turmoil in contemporary Africa is suggested as evidence that the collapse has already begun. Although we are in for a time of "great tension, with wars, famines, pestilences, and general economic breakdown of the established order, there will probably be survivors".[14] But law and order will break down and our societies will disintegrate. By 2020 a new order will be emerging, though with global "all liquids" output at 1990 levels (approximately 65 MBOPD, in contrast to 85 MBOPD today). Sub-Saharan Africa will continue through the 21st century little changed – bar a decline in its population from around 733 million to a "sustainable" 200 million. Sometime around the 22nd century, however, the world will magically reach a new equilibrium with nature. Humanity can look forward, eventually, to a "New World". But the intervening transition and crisis will be rough, rude and even brutal.

With the upsurge of writings on Peak Oil has come the formation of ASPO, a spate of conferences, local groups formed in cities and universities, international media coverage, Peak Oil websites and blogs. The Peak Oil community is ideal for viral marketing of its concerns and has grown considerably in numbers. It forms a sort of self-perpetuating community.

ASPO holds an annual conference, the most recent (at the time of writing) being that held in Pisa, Italy, in July 2006.[15] Among the speakers were many of the leading lights of the Peak Oil lobby: Jean Laherrère (ASPO-France), Kjell Alekett (president, ASPO International), Chris Skrebowski (ODAC and editor of *Petroleum Review*), Richard Heinberg (formerly of the Post Carbon Institute, now promoting the Oil Depletion Protocol) and many others.

Delegates consisted of all kinds of groups and individuals: retired geologists, practising or former bankers, social scientists from academia, electrotechnical engineers, policy advocates, assorted minor politicians, NGO groupies, former and now-retired executives from Total and a few other oil companies, authors of Peak Oil books, consultants on Peak Oil mitigation strategies, theoretical physicists, wind experts, environmentalists, physical chemists and chemical engineers, electric vehicle specialists, cybernetic technicians, solar energy interests, Kyoto climate-change advocates, oil futures product traders, mathematicians, technologists, ecologists, political scientists, film-makers, local town councillors, biologists, journalists, civil engineers, sustainable development theorists, third-world development advocates, self-proclaimed energy experts, research types, peak geeks and others besides. There was even one amusing observer who had long traded in cockroach and pest eradication on world oil platforms ("call me Cockroach Dundee").

Some groups, however, were represented in minimal numbers, namely senior corporate oil players, upstream investors, portfolio managers, industry strategists, exploration and new ventures managers, active geoscientists from the operating corporate oil world, oil economists and serious investment players from the oil markets. Yet some such types had been attendees in the past. The numerous comments made on companies, economists, certain international institutions and the like were often severe and critical.

Unsurprisingly, given the diversity of participants, there was a range of opinion on display. Yet on certain core themes there was, broadly, a consensus. Oil production either has already peaked or soon will (say around 2010) – to be followed, some commentators added, by not too distant peaks for gas and coal. Not even all of today's mega-project capacities in oil, with their slippage and problems, would avert the coming crash. Society remained complacent, either fooled by self-inflicted myopia or manipulated by the politicians and elites. Governments, especially Middle Eastern sheikhdoms (most notably in Saudi Arabia), and international energy agencies such as the USGS, IEA and EIA issued false oil market projections and dodgy estimates of oil reserves. To the list of miscreants, the faithful added BP (especially its CEO, Lord John Browne), with what some saw as its flawed "Bible" of worldwide data, economists (especially the "flat-earth" variety) and the deeply suspect optimistic number-crunchers found in consultancy groups such as Cambridge Energy Research Associates (CERA)-IHS Energy. Some even believed that the wars now afoot have been designed to leave scarce oil in the ground for the victors to exploit and that the then high oil prices were merely a smokescreen for profiteering. In short, says the ASPO consensus, do not believe what you read: the fact is, we're heading for a smash.

A typical contribution was that from Chris Skrebowski, who warned that crude supply to come from 68 or so mega-projects worldwide is not capable of producing enough oil to meet future world demand. Such factors as project slippage, over-optimism by companies, or slower enhanced oil recovery (EOR) results could accentuate the problem. Depletion will be the mistress of our future distress. There will be an unbridgeable gap between supply and demand. Oil will peak in the first quarter of 2011 at a world level of 92–94 MMBOPD.

What can save the world now? One answer might be a greater appreciation of economics – but little of this was on display at the ASPO conference, where the minimisation of matters central to economic theory has apparently become a leitmotif of purity inside Peak Oil. Instead the world must seek salvation by weaning itself off its addiction to oil. We must all learn to forgo our consumer world and its hydrocarbon foundations. The *cri de coeur* of the conference was that all must live simply in order that all might simply live.

For salvation, ASPO relies in particular on its Oil Depletion Protocol. This is its tool for "avert[ing] oil wars, terrorism and economic collapse".[16] It is based on the assumption that the world needs to wean itself off its addiction to oil. The protocol requires that no country produce oil at above its current depletion rate and that oil consumption be reduced by the rate at which the world's oil is becoming depleted (say, 2.6% per year). It is proposed that it be restricted to conventional oil (with a cut-off at 17.5 degrees API) and should exclude oil sands, polar oil and deepwater oil.

The system envisaged in the protocol would require a huge number of import quotas (which may, if necessary, be auctioned off). World leaders and nation states will be solicited to sign up. Individual citizens too are invited to endorse the protocol and sign personal pledges. No incentives or penalties are mooted for adopting or rejecting the protocol: the nations of the world simply need to recognise that we will all be better off from its adoption. As a first step, countries are encouraged to conduct oil vulnerability audits.

Peak Oil adherents at the conference saw some signs of hope in this initiative. Sweden has already begun a plan to de-oil itself at the rate of 3.5% a year but for complex reasons. Ireland and Iceland are said to be considering various proposals. Several municipalities in the US are reported to be taking steps to reduce their oil vulnerability. San Francisco has passed a Peak Oil resolution. Hope is expressed that Kuwait will revise down its oil reserve estimates and in line with a Kuwaiti constitutional proposal reduce production.

The Oil Depletion Protocol is, of course, stunningly ambitious. Can we really imagine world leaders such as Vladimir Putin and Hugo Chavez, indeed any OPEC oil producer, rushing to sign such a protocol? Quite apart from such doubts, the sheer scale of the project is overwhelming. At present, the invisible hand of the imperfect oil market co-ordinates through its automatic price mechanisms the daily tactical and strategic decisions of governments (over 200 of them), multiple oil and energy authorities, thousands of oil and energy companies, countless worldwide investors and exploration interests, and billions of oil consumers. A global system of oil rationing, aimed at stabilising the price, would require a new Gosplan-style plan and bureaucracy for the entire world. It's a most unlikely idea.

Much of ASPO's energy, of course, is devoted to the business of forecasting – both devising its own forecasts and critiquing others. Here one key figure is Jean H. Laherrère, whose forecasts, based on a parabolic fractal model, are ASPO's most mathematical and sophisticated.[17] Noting that the data available (for instance, IHS and Wood Mackenzie data) are considered flawed, Laherrère now forecasts the date for the moment of Peak Oil within a broad range of 2012–20. This, it will be recalled, is later than the dates Campbell forecasts, now varying around 2006–10. A stretching of the consensual date seems to be arising within Peak Oil.

The fact that the Peak Oil date emerges from these forecasts as a rapidly moving target is no minor matter. Laherrère is, after all, a leading Peak Oil theorist ("All ASPO is based on Laherrère," someone said at the conference). Other forecasts, though broadly in line with the overall import of ASPO's message, introduce further latitude. Former Total executive Pierre-René Bauquis, much less aligned to ASPO's central depletion model, estimates that oil production may peak at 100 MMBOPD (+/-5 MMBOPD) in 2020 (+/-5 years). The Institut Français du Pétrole (IFP) forecasts that the peak will come in the range 2015–30. So on the edges of ASPO is found greater elasticity in timing for the oil peak.

Forecasting is always a difficult business. Perhaps one should never issue a forecast date within the range of one's life expectancy. An Arab proverb springs to mind: "He who forecasts lies, even when he tells the truth." But it is difficult to see why, in that case, the IEA's forecasts and those of others in contrast should be greeted with such suspicion and, indeed, accusations of deception.

The shifting back of the chosen peak dates and the broadening of the range has one big advantage: it makes the forecasts more likely to coincide or approximate with the eventual truth. So Peak Oil somewhere within 2006–30 seems a fair spread. It means that, at the time of writing, the peak could even on these estimates lie a whole generation away – which leaves, the sceptic should note, plenty of time for yet further revision.

The role of other hydrocarbons besides oil was the subject of some speculation at Pisa. Some commentators foresee a period in which coal will become dominant and coal-to-liquids (CTL) will become a major issue. There might also be a boom in stranded gas and liquefied natural gas (LNG). It would seem, however, that absolutely everything will ultimately peak. The forecasts for such energy sources take the same parametric form and rely on the same kinds of assumptions as those made for oil. Jean-Marie Bourdaire (formerly of Total) even forecasts a triad of "GDP-killers": peak oil in 2017; peak gas in 2030; peak coal in 2040–50. This triad appears as a series of energy-driven economic catastrophes.

Even world population will peak, according to one source, at around 7 billion souls in 2050. Louis Arnoux offers calculations on hydrocarbons in general based on a per head basis. According to him, barrel-of-oil equivalent (BOE) per head already peaked globally in 1979. World supply is now only 7 BOE/head and will fall to 4 BOE/head by 2050; by 2100 it will be below 1 BOE/head.[18] This is a vision of not only the end of oil but of all fossil fuels: oil, gas and coal. It links hydrocarbon decline to planetary overload.

The focus on the business of forecasting peaks was supplemented by discussion on broader aspects of the social context. One commentator, Luigi De Marchi, sources the psychological roots of the energy crisis, focusing on the population explosion occasioned by our failure to confront multiple sexual taboos (think here philosophies within the Catholic Church, Latin American masochism, dogmatic Islamic cultures, and the like).[19] It appears that there are just too many people. No serious premature peaking there, it seems. The Peak Oil economic opportunity lobby, however, chose to concentrate on the business opportunities offered by the peaking of oil production as a silver lining in a darkening and clouded world.

Peak Oil theory is supplemented too by concern over the so-called "Military–Oil Complex" and the accompanying geopolitical scenarios. Commentators point out that a US–Iran confrontation could seriously disrupt supplies. Anglo-American domination is threatened, it is said, with the result that NATO increasingly has to operate "out of area" in an attempt to secure the reserves in Central Asia, the Persian Gulf and even Russia. Almost any event, it seems, may be explained by this new irreversible scarcity of oil: the Afghan and Balkan wars were seen as part of a strategy to encircle Russian reserves; the independence of Timor Leste (East Timor) was designed solely to remove offshore oil from Indonesian control; the conflicts in Somalia and Yemen are in essence about the control of the strategic routes through (and possibly fields in and around) the Red Sea and the Horn of Africa; and US assistance to Bogotá is driven by the need to protect Colombian oil production and provide a platform for securing control of Venezuelan oil in due course. The importance of all of these manoeuvres is heightened by the West's need to outflank Chinese oil companies.

ASPO's view of the contemporary world is supplemented by its own, largely geologically determined, version of history and the future. America's overseas policies are all about oil and its finite scarcity. The institution of the nation state has started to fracture because of oil conflict only (witness the balkanisation of contemporary history) and will continue to do so. There is also a strong undercurrent of anti-globalisation (think Porto Allegre here, not Davos) in ASPO thinking, given preferences exhibited for the shrinking modern economies to cope with the coming oil crisis and some desire shown for small-scale communities to replace our obsolescent world.

Among the more wide-ranging discussion, some commentators even considered issues usually regarded by Peak Oil theory as irrelevant, namely economics in general and the price mechanism in particular. But their thrust and conclusions were little different. Renato Gosso provided forecasts of Peak Oil production (2007) and depletion (90% by 2019) in a diffusion model incorporating the effects of changes in price and technology.[20] Neither would rescue the world.

Mamdouh Salameh (ex-UNIDO and former World Bank

consultant) attributed the high price of crude (around 2006) to the fact that production had peaked, apparently, in 2004.[21] A speaker from Ecohabitat and the Pakistan Wind Energy Program argued that the market could not solve the problem posed by Peak Oil: high oil prices "do not decrease economic growth", we are advised, unless $90 per barrel is reached.[22] Another speaker, however, forecast that the price for Arabian Light could reach $200 per barrel (at 2000 prices) over the period 2012–20. Some felt that such price rises would trigger the development of synfuels and the expansion of new-generation nuclear energy. Yet the potential role of crude price cycles, technologies, new fuel forms and substitution effects were not viewed by most as likely to undermine the vivid scenarios that accompany Peak Oil forecasts.

Given the nature of the Peak Oil debate as a whole, it is not surprising that it has attracted increased attention from politicos. For example, Michael Meacher, a British MP and former minister for the environment, entered the debate with gusto. As the oil runs out, according to Meacher, "the economic and social dislocation will be unprecedented ... and the political significance of this is almost incalculable".[23] (Peak Oil, has not, we note, calculated quite everything.) The best option for the future, apparently, is the inevitable "New Energy World Order" based on the infinite supply of renewables "to power the world economy". The problem for Peak Oil, however, is that it has yet to attract the support of big hitters in politics. As a result, the political debate sinks to the level of the lowest common denominator.

As well as the politicos Pisa attracted many journalists and writers, for whom ASPO provided ideal copy for dramatic

editorials, as well as a convivial couple of days in the brilliant Tuscan sunshine. The writers included Sabina Morandi, an Italian novelist, whose works include *Petrolio in Paradiso*, a novel of oil and ecology in conflict in the jungles of Latin America. Sadly, one finds on inquiry that it lacks the key ingredient (surely) for any novel about peaking: petro-sex.[24]

One constituency almost wholly absent from Pisa 2006 was corporate oil. No consensual view on Peak Oil has emerged from the core of the industry. ExxonMobil apparently sees no imminent peak (or at least has not declared one). Chevron has alluded obliquely to corporate concern on future oil supply (perhaps just their own) and also worries about private companies' access to oil reserves. Any statements from BP tend to be regarded by Peak Oil adherents as "beyond probability". Total (France) has generally not set any date although it is well aware of the issues. Shell meanwhile sees the world as one full of oil, gas and energy resources. State oil companies appear even less concerned, as confirmed by views expressed to the author by a range of national oil companies.

The lack of corporate engagement with Peak Oil may in part be because of other experience in the 1970s–80s as well as disagreement with its core theses and model. Then too debate focused on forecasting (though in those days of crude prices rather than oil production peaks). The general inaccuracy of price forecasts then has perhaps engendered a sceptical approach on the part of corporate oil now. For exploration players and independents, we should note, forecasts of oil scarcity promise a world of new opportunity accompanied by higher margins and enhanced capital flows. Such optimism is not shared by Peak Oil theorists, for whom the potential of exploration and new plays is considered strictly limited.

No account of ASPO and its conference would be complete without mention of its band of sceptics and even mild heretics. Many of the former have attended several conferences over the years. Some perhaps even show signs of Stockholm syndrome: they have, as it were, begun to deploy the arguments and defences of the intellectual captors. In general, however, the sceptics were a source of some astute, if not particularly original, realism. They were well aware of increases in reserve estimates and oil recovery rates. They pointed out that intertemporal comparisons of companies' oil production are skewed by the incursion of resource nationalism. They recognised that traded global oil profiles will shift in response to any change in global oil production. They acknowledge too that the displacement of corporate oil by state players introduces slack into the upstream industry by substituting less efficient asset management (and, especially, reducing commitment to exploration).

Among the milder sceptics is Robert Hirsch. While accepting that "peaking will some day appear", he criticises some of the thinking on which many forecasts are based. Hirsch stands out, in particular, for considering the effects of economic mechanisms, arguing that a number of possibilities could mitigate any supply deficit in crude oil. He draws attention to the potential of coal-to-liquids, gas-to-liquids, heavy oils, shale oil, commercial ethanol, biomass-to-liquids and improvements in vehicle fuel efficiency.[25] Moreover, oil-energy intensity has improved (even if the record remains uneven) and more could readily be achieved. In addition, changes in tax subsidies could correct much inefficient oil usage through the repricing of fuel. If a crash mitigation programme were initiated, fuel shifts could in 15 years generate around 25 MMBOPD of new liquids supply. This would be significant.

The Peak Oil church is perhaps now broader than it once was. There remain at its core its faithful apostles and canonical texts. And there is a fringe with overheated imaginations concerning Peak Oil's social, political and demographic scenarios. But there is also the emergence of what might be called Peak Oil Mark II, comprising sceptics, dissidents and heretics (including some discussed). For them a reconsideration of oil reserves, flow relationships and the impacts of government policy and economics are not unthinkable. This group includes what might be called the "peak adaptors": those who respond to predictions of oil peaks by pointing to the potential of the energy spectrum as a whole (including, most heretically, nuclear power). Peak Oil Mark II may even be starting to make good the movement's failure to engage the mainstream industry, though the degree to which dialogue is welcomed by Peak Oil purists is open to question.

Across the movement as a whole, however, there remains a powerful predilection for pessimism. While the outside world continued as always, the Pisa conference was permeated by empirical gloom. Even so, it looked like everyone was still having a fine time.

4

Classic "end-of-oil" syndromes

An event such as the Pisa conference leaves one with a kaleidoscopic mix of ideas about the peaking of oil – its existence, date, shape, scalar impact, causes, inevitability, unintended consequences, societal implications and meaning. Is there, one wonders, some key to the meaning of it all – some Peak Oil equivalent to the Da Vinci code? Or is this all just a manifestation of what Bryant Urstadt suggests in *Harper's Magazine*, where the search for Peak Oil's political soul finds it reflecting scenes from the liberal or post-liberal apocalypse?[26]

The Peak Oil movement has now spawned many new offshoots, some apparently not unconnected to the Four Horsemen of the Apocalypse. Groups now meet worldwide – even in oil-soaked Ohio. In New York City the Peak Oil Meetup Group (one of 39 such groups across the world) meets monthly, providing a sort of group therapy to help its 340 members cope with the coming end to our industrial civilisation. Its members feel neglected by society, cast adrift by the politicians, shocked at ever hearing themselves described as "crazy".

Urstadt has revealed the thoughts of its members on many

aspects of the nightmare ahead of us that will arise from Peak Oil – from the onset of fascism to questions of how to turn asphalt into gardens and whether to cover skyscrapers with solar panels or place windmills on their roofs. How, they wonder, can we create self-sufficient "lifeboat" communities to ride out the coming storm until sanity returns to the world? What can we do to cope with the simultaneous impact of oil depletion and the coming of a new ice age ushered in by global warming? What supplies should we store away? Should we buy gold or silver coins to hedge against the coming financial collapse? Where is there to seek refuge, other than the warmer climes that will be inhabited by nutty born-again Christians and gun-toting southerners? Their angst runs deep.

The New York meetings brim with information, statistics and cross-disciplinary interfaces. From it all an observer could easily be left with the impression that disaster is upon us now. There is a back-to-the-land ideology (and indeed plans for putting it into practice), supported by a mixed ethic of altruism and self-preservation. Faced with a forthcoming mass exodus from cities we will need to grapple with the complexities of establishing eco-villages. How should the strong survivors cope with the weak? What will happen to the elderly? Plans are made for the management of violence and social breakdown in the tough survivalist world to come.

Though the group includes some right-wing members, the movement in general has, according to Urstadt, all the hallmarks of the "liberal left behind". It is a movement apart and aside from America's dominant conservative politics. It is tempting to see in all this a psychosis with its roots in the long tradition of apocalyptic thought in American Puritanism. Indeed, there are echoes of a variety of traditions in American thought: millenarians making their survival plans; occult groups searching for final redemption; even the alien encounter schools promising lift-off to distant planets for the chosen and faithful.

Most such quasi-cache movements profess some inside knowledge presented as scientific. Group identity is sustained by an intimacy with its particular version of "the facts" combined with a system of beliefs held with deep conviction and a disdain for establishment thinking. What gives Peak Oil thinking its unique appeal is that it combines a conviction that the moment of catastrophe can be forecast precisely with, in practice, a range of shifting peak forecasts. Enviably, it offers the psychological attraction of "pick 'n' mix" flexibility and a sense of complete certitude. There is, in effect a menu of traumas on offer – international conflict, oil wars, the disintegration of the social fabric, and the decline and fall of empires. It is enough to have a wide appeal and attract a diverse group of the faithful.

How, one wonders, did Nostradamus fail to spot it all? Nostradamus Online, however, appears to be in line.[27] According to an interpretation of *Century II, Quatrains 3 & 4* it reports that in a time of change the Antichrist, with its source of power located in the Middle East, will go nuclear – or "postal", to use that quaint Americanism – and hence western nations will seek "to diminish the threat to oil supplies", so we are informed.

All this concern arises from a lack of sustainability in our oil and hydrocarbon world. We are reminded that per head oil production (according to Arnoux) peaked 27 years ago. The option of sustainability was only ever a Club of Rome fiction: the real choice, we are told, is between a very harsh world or ultimate catastrophe.[28] Fossil-fuelled growth has smashed into our ecology like a series of tsunamis. The very survival of the human species is under threat. At its peak, the global population will number 7 billion, all of them "fossil-fuel guzzling hyper-predators". By 2100 it will have collapsed to a mere 2 billion. The harvesting of solar energy ("the greatest business opportunity since the last ice age") is apparently our only hope.[29]

Increasingly, too, corporate actions are interpreted as evidence that big business is coming to believe in Peak Oil theory. After all, when Deutsche Bank examined the end-of-fossil-fuel-hydrocarbons scenario, it took the view that the issue of scarcity in coming years and decades must be taken seriously.[30] The oil companies' behaviour is taken as witting or unwitting evidence that they share Peak Oil's convictions: ExxonMobil's fuel-efficiency drive is seen as a sign of alarm over the imminence of the oil peak, and Chevron's global public relations campaign, Will You Join Us?, as a device to deflect debate and maintain investor confidence. For corporate oil, it seems, it's doomed if you do and doomed if you don't.

Among the most dramatic scenarios that resonate within Peak Oil is that contained in the Olduvai theory, dating from the late 1980s.[31] This theory has been called "unthinkable, preposterous, absurd, dangerous and self-fulfilling" and chimes with Peak Oil theory and data for oil, energy and demographics. Industrial civilisation has, according to this thesis, a lifespan of 100 years (that is, 1930–2030). It will end with an epidemic of permanent blackouts of our high-power electricity networks. When the lights go out, "you are back in the Dark Age … and the Stone Age is just around the corner". Welcome to the Kalahari – or, to be more pertinent, Olduvai gorge.

The trajectory of our decline is, it seems, a matter of "slope, slide and cliff". Olduvai theory includes a series of eight critical events finely calibrated along a not very long timeline. The third event is the end of cheap oil, which apparently took place in 1999. The fifth event is Peak Oil (2006). Then comes the moment when OPEC accounts for more than half of the world's oil supply and nearly all of its exports. We reach the cliff in 2011, when there will be brownouts and blackouts across the world followed by the collapse of agribusiness, the supply of fertilisers, food supply, transport systems, banking, modern medicine, plastics, defence systems, construction, manufacturing, trading, shipping, and so on. (More grim details are available on the internet.[32])

Among the flood of literature now appearing on post-Peak Oil scenarios, one in particular has been much heralded: James Howard Kunstler's *Long Emergency*.[33] It has, indeed, been compared in importance to Alvin Toffler's *Future Shock*. Nature, according to Kunstler, is about to bite us back. It is not just Peak Oil that must concern us. We will suffer from climate change, water shortage and habitat destruction. In this context, hopes of smooth energy transitions are untenable: the depletion of oil will trigger decades of gruesome conflicts. The only hint of a silver lining is that although the population will die down, humanity will not entirely die off.

According to the theory of Thomas Malthus, population expansion is inevitably limited by a number of unpleasant checks (famine, for example). Malthus, according to Kunstler, was in essence correct. The orgy of consumerism that we have witnessed over the last century or so may appear to contradict this hypothesis. But that is an illusion. The living standards that we have enjoyed have been sustained only by virtue of a cheap oil supply. As oil reserves become depleted, so Malthus's hypothesis will kick in, and fast. "The journey back to non-oil population homeostasis", asserts Kunstler, "will not be pretty".

Like others, Kunstler has a theory not only about what will happen to us, but also about why we do not want to hear the truth. Corporate executives face mentally disabling pressures (not least because they believe the economists they employ). They so often end up believing their own propaganda. Politicians, in turn, are hardly likely to announce the end of the American dream or its global equivalent: no call for radical downshifting is likely to win many votes. In the circumstances the USGS merely assists its master, the US government, with its delusional, inaccurate spin.

The world, according to Kunstler, has never faced such clear and present danger. Neither the Second World War nor the Cold War was close to what lies before us. As the oil begins to run out, the boundary between politics and economics will dissolve. China's new, voracious appetite for oil will intensify the competition for remaining oil supplies. The oil-rich states of the Middle East, all in the backyard of a resurgent Iran intent on arming itself with nuclear weapons, will become the chief locus of global conflict. In this Last World War, it will be global anarchy, rather than the US, that prevails. There will be no salvation through the deliverance of alternative energy supplies: our age-old belief in deliverance is merely a form of magic, comparable to a cargo cult.

The predictions that emerge from perspectives such as the Olduvai theory and *The Long Emergency* hinge on the symbiotic marriage between the key Peak Oil assumptions – that we know what lies below the ground and that we can infer what happens above the ground as a result. Its components are the constraints of geology ordained by nature; the fragility and subservience of human society before omnipotent power; evil acts (the consumption of the forbidden fruit of crude) calling forth eternal damnation; the dire prophecies and the coming Apocalypse; the preordained cycle from ashes to ashes (a non-carbon-neutral image if any); and, with luck, an eventual harmony for the lucky few. There is discernible in such images a sort of Peak Oil theology.

5

Peak Oil theoreticians, advocates and acolytes

The Peak Oil movement is if nothing else multifaceted. It includes ideas and proposed solutions, websites and blogs, predictions and postulations, groups and individuals, protagonists and defendants, techno-geeks and populists, overinflated politicians and activist acolytes, single-issue lobbies, various ideologues, defensive and offensive strategists, institutions and targets of wrath, conviction and/or bandwagon journalists and media gurus, old hands and the newly initiated, authors and film-makers along for the ride, new celebrities and assorted wannabes.

There is, of course, no formal qualification for joining the movement, no regulatory body, no institutional membership. Individuals may associate with ASPO and there are other, less formal, ways in which they can affiliate to the movement – signing Peak Oil testaments, for example, or linking to websites. Though membership of the movement confers no direct benefits, it does offer a feeling of being part of an intellectual elite and vanguard for global change.

The Peak Oil movement has by now developed a lengthy

history. As we have seen, it was half a century ago that Hubbert developed his Peak Oil idea and method. The first Peak Oil website is said to have appeared around 1980, in the very early days of the web. Once regarded as part of the lunatic fringe, the movement is now seen more as a respectable if sometimes misguided special interest group. Perhaps this change in perception is in part a result of the public's ignorance of the complexities of the oil industry which makes it a soft target. But it also results from the movement's energetic use of the internet and its skilful, sustained, public relations campaign. At the time of writing, a search on Google reveals over 23 million relevant items. The numbers are growing fast and the movement is now well enough established to outlive its founding fathers.

One of the major publications in the Peak Oil market has been *The End of Oil*, written by Paul Roberts (a journalist) and published in 2004.[34] His starting point was the high and presumed dangerous level of water cut in the Saudi super-giant field of Ghawar, the world's largest. In a flash of insight Roberts took this as a metaphor for the state of the oil industry in general. His dramatic and informative work presents a colourful narrative that includes the end of easy oil, the control of crude oil tied inexorably to war (including the first Gulf War), Saudi mendacity, US strategy driven by the need to acquire oil reserves, the major corporations ridden with supply anxieties, growing oil-energy insecurity, the development of alternatives to replace a depleting oil stock and to create the next energy economy – all set on a planet that is becoming "too hot". And yet for all its drama, *The End of Oil* offers one of the more moderate scenarios.

The same cannot be said of former Los Angeles policeman Michael Ruppert. His view, evidently, is that the Peak Oil movement has been far too complacent (see his website www.fromthewilderness.com) and needs to become activist. Ruppert itemises oil disasters that have damaged supplies and potential shocks that would make the situation worse. We learn of potential earthquakes in key oil-producer zones, global warming impacts, aggravated civil unrest in oil areas, attacks on Saudi Arabia's facilities, conflicts in Mexico and the Caspian, and so on. America is particularly vulnerable because of its car culture, customary cheap fuel, limited public transport systems, intensive and energy-reliant home ownership, and undiluted consumerism. About all this there has been, apparently, a conspiracy of silence. The mainstream media are in denial. The public is led to believe that contemporary world conflicts are not about oil. An array of policy options and a trust in the goddess of technology diverts them from seeing the flaws in the system.

In order to survive, we must abandon our luxuries. Some hope is seen in the Oil Depletion Protocol and also in the Post Carbon Institute, which has 90 affiliated groups in North America – though apparently 9,000 at least are needed worldwide. The true heroes of the age are those who are moving into the post-oil paradigm with low carbon footprints. It all sounds a little like what Chairman Mao's Great Leap Forward was supposed to be – a radical, once-and-forever transformation.

The weakness of Peak Oil theorists here, apparently, is that they have not focused on the transition between our present and the future paradigm. They stand accused of building not lifeboats, but alternative luxury cruisers, for rescuing the world. It reminds us of the image of Noah's Ark: it is as if in the imagination the forthcoming oil drought acts as a *fin du monde* inversion of the Biblical deluge. We need, it seems, not merely to make projections about the world, but urgently to disengage from the current global economic paradigm.[35] In true American revivalist fashion, Ruppert, the latent politician, ends with the promise of "a new understanding of God awaiting those who survive".

Many political activists of long pedigree are now involved in some way with Peak Oil. Paul Ehrlich, famous since the 1960s for futuristic projections, has mounted the bandwagon.[36] Peak Oil has clearly acquired traction with the old left, as well as with members of the intellectual glitterati such as George Monbiot, a well-known journalist. Despite offering the heretical view that "no one has the faintest idea"[37] when oil production will begin to decline, Monbiot claims that the people sitting on the world's reserves are "liars", and that Bjorn Lomborg (who in *The Skeptical Environmentalist* provides an optimistic view) is just plain wrong.[38] Peak Oil also attracts the supporters of solar and renewable energy. The energy game is, after all, a competitive one: anticipation of shrinkage in the oil supply provides a heaven-sent opportunity for non-oil lobbies.

It is useful in this context to examine the contribution of Jeremy Leggett. By his own account, Leggett, who has an earth-science pedigree, was in his years as a lecturer at Imperial College (London) a "creature of Big Oil". Since then he has become involved with Greenpeace and the non-oil commercial energy world. His book *Half Gone* engages fully with the Peak

Oil debate. It covers the usual issues – oil discovery deficiencies, depletion dilemmas, the weaknesses of technologies, problems in unconventional oil extraction, and so on – with the usual paraphernalia (Hubbert's bell-shaped curves, criticism of BP's statistics, and so on). Factors such as geopolitical constraints and historic industrial underinvestment are viewed as marginal; the central debate is presented as that between those who argue for an early date for the moment of Peak Oil and those who forecast a later one.

All this might sound like just so much conventional Peak Oil wisdom. But Leggett provides us with a dramatic twist that we might describe as "oil depletion meets global warming". According to Leggett, the depletion of oil will inspire a coal rush, with potentially disastrous consequences for the environment. But maybe, just maybe, competition from renewable energy, especially solarisation, will save the day. The key concept here is that of the tipping point: that a new idea can, with enough support from an enlightened elite, tip over into tomorrow's new paradigm.

Another key activist is Richard Heinberg of the Post Carbon Institute, author of *The Party's Over: Oil, War and the Fate of Industrial Societies*.[39] Heinberg's contribution to the debate is certainly original in parts. Drawing on a now declassified report, he reveals that in 1977 the CIA forecast that Soviet oil production would peak in the 1980s. This event was forestalled only by the demise of the Soviet empire and the crash of the Russian economy.[40] The Soviet Union was apparently consigned to history by connivance between the US and Saudi Arabia to flood the world market with cheap oil in the mid-1980s. What intrigues Heinberg is the CIA's insight into the Peak Oil issue and, according to him, its use as a key plank of Cold War strategy. The CIA, says Heinberg, "must therefore understand the issue of global peak oil". Does the CIA, it may be asked, harbour a secret strategy for dealing with this impending global event? It's always delightful to be able to involve the spooks.

Peak Oil has become a subject of deliberation in the US Congress. The resolution from Representative Roscoe Bartlett of Maryland to address the "inevitable challenges of Peak Oil" attracted some 14 like-minded members. In December 2005 the House Subcommittee on Energy and Air Quality heard evidence from, among others, Dr Robert Hirsch, ASPO's president Kjell Alekett and CERA's Robert Esser, as well as Ron Cooke, a self-styled cultural economist.[41] Bartlett dazzled the political luminaries on the committee with a discourse on bell curves, peaks here and there, inexorable rising oil demand, the dangers of opening the Alaska National Wildlife Refuge (ANWR) in Alaska, Canadian tar sand gas-guzzling, deification of the market, mixed with what sounded like straightforward realism. ("I want to be 60," Bartlett said with grave emphasis.)[42]

The testimony put before the subcommittee was decidedly mixed. Hirsch, an engineer-physicist, argued, in line with Peak Oil theory, that the price-reserve nexus "was not understood by economists". This seems a stretch. According to Esser (a geologist), however, Peak Oil is "a not very helpful concept". He explained that CERA's proprietary world oilfield database had recently been re-audited, confirming an "enormous scale of field reserve upgrades of existing fields". In effect the world was not running out of oil. CERA could find no evidence of a peak before 2020 and envisaged a long and undulating plateau between two and four decades from now.

The cultural economist in turn produced a long critique of CERA's heretical view. According to him, an imminent oil peak could be avoided only if certain multiple preconditions were met. To be specific: peace in Iraq; international political stability; the containment of terrorism's threat to the supply chain; OPEC's reserve estimates turning out to be accurate; modest future depletion rates; oil production for most projects remaining on schedule; increasing capacity in the Middle East; a ratio of sustained EROEI (energy return on energy invested) of greater than 1; increases in unconventional oil output; adequate oil/energy infrastructure; the granting of permits for non-Muslim technicians to work in the Middle East, North Africa and Caspian countries; capital sufficiency for oil projects; decreasing rates of growth in the demand for oil; a retreat from resource nationalism; beneficial impacts from new technology advances. It's possible of course to generate almost any number of such ex-ante conditions. The most pertinent point, however, is that in contrast to conventional Peak Oil wisdom, the cultural economist's list emphasised the importance of many economic and above-the-ground factors. Perhaps the answers do not all lie in the soil.

It will be seen from these various synopses that many commentators on Peak Oil scenarios tend towards dramatic or in some cases extreme stances. However, Peter Tertzakian in his book *A Thousand Barrels a Second* at least claims to be taking the middle ground between, on the one hand, the usual "threat scenarios" and, on the other, the "extreme view" that there is absolutely no problem in world oil supply.[43] It is not entirely clear who would subscribe fully to the latter view: certainly nobody with firsthand knowledge of the upstream industry would see the world as a perfect playground offering uninhibited supply, unrestricted reserve access and zero oil complexity.

Tertzakian believes that we might in future encounter Peak Oil problems. But he points out that the evidence of history is that energy sources come and go. There is a replacement cycle in which, as established fuels reach breaking point, alternatives emerge. He argues that the challenge provided by an approaching breakpoint in the supply of light sweet crude will need to be met through both market adaptation and government intervention. Solutions will vary from country to country but will involve technology development and interfuel price changes. Responses will include improving energy efficiency and conservation, the development of renewables, a switch to fuel from oil sands projects and lifestyle changes. Past initiatives in oil rebalancing help: the better the energy mix today, the easier will be the transitions of tomorrow. Overall, Tertzakian's vision is one of a world responding to a difficult challenge by muddling through. It has been like this before: it may be so again.

Let us finish this survey of post-peak scenarios, however, not on a note of muddling through, but instead with one that may be Peak Oil's real McCoy. *The Final Energy Crisis* is a collection of 23 essays, many of them written by Andrew McKillop (though there are around a dozen contributors in all).[44] The book provides "detailed, irrefutable scientific evidence" for its resolutely Peak Oilist thesis. Apocalypse is carefully scheduled for 2035. By then, oil production will have plummeted to 25% of today's levels. Nuclear energy and technology development will not be able to make good the shortfall. Africa will suffer a continental-scale meltdown and be subject to Belsen economics. As petro-capitalism collapses the Anglo-Saxon hegemony will give way to Asian dominance. A series of wars (perhaps nuclear) will be fought over oil and gas. It's Armageddon.

Under the twin challenges of climate change (the Kyoto initiative having proved unworkable and carbon sequestration having failed) and the disappearance of the fertiliser industry, our current agro-business is doomed. Solar and renewable energy will not be able to support today's population and living standards. As a result we face the grim prospects of food wars and population decline. We will have no alternative but to decompress our energy-intensive but flawed societies. Only agrarian communities based on small farms and local food will be sustainable (we hear here the echo of *Small is Beautiful*). Humanity will evolve backwards from complexity to simplicity – the first such recidivist species in history.

6

The ultimate cult of concern

From the point of view of Peak oil theory, oil seems to be the *fons et origo* for everything of value about our civilisation: the production of oil has made possible our achievements, so its depletion will lead to our downfall. Fortunately, from this point of view, we at least have ASPO as our prophet to guide us. So far we have viewed ASPO largely through a snapshot – that of the 2006 Pisa conference. Now let us stand back and look at this prophetic body more in the round.

ASPO-1, the initial annual conference, was held in 2002. It concerned itself with announcing the daunting problems facing us.[45] This was perhaps Peak Oil's coming-out party, its arrival on the world stage, even though the content of the event built on much intellectual thought that had done before. Even so, here ASPO found its legs and it has made swifter running since.

ASPO-2 was held in Paris in 2003. The conference attracted wider interest and also subsidy from Total, IFP and Schlumberger. Since then these companies have ceased to act as sponsors. One may speculate as to why: maybe their optimism clashed with the pessimism that characterises ASPO thinking; maybe, suggest some ASPO supporters, they are reluctant to face their own dilemmas; or maybe they wish to distance themselves from the geopolitical side of ASPO debate. Certainly the BBC's documentary *The War for Oil* went down a treat and resonated through the conference. In any event the tradition of an annual meeting was established.

ASPO-3 was held in Berlin in 2004. By then the participants numbered around 250. The conference attracted the growing interest of the mainstream media, thanks in part to the debacle concerning Shell's reserve estimates. BP, ExxonMobil and the IEA were among those involved as participants – not altogether harmoniously. The IEA's Dr Fatih Birol, in particular, came under criticism, and the integrity of the IEA case was called into question. Thanks in no small measure to the contribution from Dr Ali Bakhtiari (senior expert, National Iranian Oil Company), the 2003 event played a role in disseminating the suspicion that Saudi oil reserves are falsely estimated and that their own peak moment was imminent. Matthew Simmons, whose *Twilight in the Desert* (2005) has become so widely quoted, visited Saudi Arabia that year. Bakhtiari ("The Prophet Ali", as Ruppert named him) had other warnings too, among them the fact that "World War III has already started". The 2003 event was also important for the development of what has become known as the Oil Depletion Protocol. It seems by this stage that ASPO was leaning towards the more dramatic and more issues at least were on the table.

ASPO-4 (Denver, 2005) attracted 500 participants and showed that interest from the media was still growing. The city's mayor gave his backing to the event. There was a growing sense among its members that ASPO was becoming established as an essential forum for debate on the future of the world. Though Professor Albert Bartlett, billed as the "Godfather of Sustainability", was honoured with an award, Julian Darley (author of *High Noon for Natural Gas*) struck a chord with his argument that "sustainable growth" is an oxymoron, as did Henry Groppe (Groppe, Long & Little) with his advocacy for crash programmes for reducing the demand for oil.[46] Simmons was undoubtedly a star of the event. His view on, *inter alia*, Saudi Arabian reserve capacity and production dilemmas has made him an iconic figure in the Peak Oil movement.

Against this background, the proceedings of ASPO-5 in Pisa in 2006 added little that was really new. There was, if anything, an even greater emphasis on the supposed deficiency of Saudi reserves and Simmons' related thesis that the world does not have a Plan B.[47] Perhaps too there was a sharper focus on the "bad guys" and those in denial – corporate oil (especially but not only BP), critics of ASPO such as economists and CERA-IHS, and various established international institutions. There was markedly less support for the event from oil companies. Corporate oil was notable by its absence. ASPO-6 was expected to be staged in Ireland in 2007.

Let us conclude this brief portrait of the ASPO movement by taking stock of this social phenomenon. When, post-Pisa, we stand a back a little, what picture emerges?

First, there is the rapid growth of the phenomenon in serial events, more participants and wider exposure. The public relations impact has grown significantly, in particular the importance it is given in the media (as opposed to within corporate oil). The ASPO organisation has grown and has bred affiliates around the world, as well as support groups and advocates in a wider range of endeavours.

Second, in the public perception of geological debate ASPO has established a bridgehead as a kind of "discipline within a discipline", in some respects monopolising wider ground than would be justified by a full-range perception of contemporary geology and its different ideas and many thousands of practitioners around the world. What can be seen is that there is no organisation that exists as a counterpoint and systematically answers Peak Oil's geoscience. In this manner, Peak Oil dominates a landscape in which the public may come to believe its conventional wisdom everywhere within the technical world. Moreover, it asserts that its view of the technical side of world oil is the only valid one.

Third, for obscure reasons Peak Oil has also portrayed itself, of course, as an enemy of the discipline of economics. This has stimulated an intellectual discourse of some acrimony between this restricted geology view and the mainstream within economics.

Fourth, Peak Oil has made itself, simultaneously, into a notable outlet for global angst and a pinnacle of certitude. This paradox is a curious and flexible one. The angst can augment in correlation to the arrival of the Peak Oil date while the forecast date can be forever put off by revision and new projections on its evolving future.

Lastly, there is the importance of the dual phenomena – the vision of knowledge and certitude about the geological and social endgame. To become accepted as a true believer, one perhaps needs to sign up to both. Not everyone does so. Our future, it seems, lies in a few Noah's Arks filled with a smorgasbord of sociopolitical remedies antithetical to society but necessary to support the lucky survivors.

Part 3

KNOWLEDGE AND MINDSETS
IN THE OIL GAME

We have seen that the dramatic social model of Peak Oil relates to specific history, various traditions of thought and contexts of perception. The most prominent source of such ideas connects to the Peak Oil theory itself. This body of thought portrays a knowledge that it asserts as broadly incontestable. It represents a mindset in a selective part of the oil world. And it holds sway in significant parts of the public mind in several modern societies.

Peak Oil's vision of PetroApocalypse derives from a particular "way of seeing" that is founded on a technical view of the oil world and its relation to geology. On these foundations have been built models intended to exclude numerous forces that shape the world's oil future.

Questions need to be asked here. Are such models the best proximate framework for understanding the oil world? What, in particular, are the implications of non-linear thinking for this technical paradigm? Is the particular geological mindset embodied in Peak Oil theory shared by the world's oil companies and geoscientists? And is oil really so different that special laws apply to it alone?

We need also to probe the depths of the theory advanced and on which the claims of Peak Oil are made. There are many different ways to handle the numbers and ideas on which the theory is based. Their model is not the only one that might be constructed.

To understand Peak Oil theory and evidence – and the intensity with which it is proclaimed – we need to provide a broader perspective. The model needs to be probed and examined in all its facets. Its choice of bounded parameters deserves scrutiny. Its various strands of thinking should be evaluated. The numbers used need attention to see if they hold up to realities encountered in the oil game. And in particular, we need to consider the views of the critics.

7

Models and non-linear oil worlds

It has taken a long time for the earliest lessons learnt by man to resurface in modern intellectual thought. In the history of the physical sciences, and even of economics, much modelling and understanding has been parametric and linear, both partial and bounded – sometimes, indeed, unifactoral. Now new ways of thinking are becoming more common. Since the late 20th century, ideas of chaos and complexity have, in effect, found their way back into our thinking and have become incorporated into our formal intellectual theories. They provide a new and enlightening set of optics in disciplines as diverse as, for example, science, art, economics and architecture.

Such revolutions in thought do not occur smoothly. New ways of thinking encounter resistance from the old. Professional bodies are prone initially to dismiss the pioneers of new thinking as heretics. It takes time for what an American philosopher, Thomas Kuhn, has called a "paradigm shift" to occur.[48] Consequently, some of the most significant notions in non-linear theory and complexity have yet to find their way into the intellectual mainstream. This is certainly the case in economics.

Once the paradigm for economics was what we now know as classical theory. That in turn was replaced, first by neoclassical theory, within which emerged econometrics, the latter seeking to provide measurements of economic phenomena based on both equilibrium and partial equilibrium models.[49] These models still dominate much of the tradition in economics. Intense intellectual battles over paradigm shifts have been fought within the learned journals, professional organisations and academic hierarchies. Only recently has an appreciation of the concepts of uncertainty, complexity and non-linear thinking begun to emerge.

We may use the term "chaology" as a shorthand rubric for this richer, dynamic and non-parametric way of seeing the world. If there is one place in which such ideas should have the widest application, it is the upstream oil world. In this light, let us consider the prism through which Peak Oil theory views the world – and the claims that it makes about what it sees there. Is there really, as Peak Oil claims, a unique, unidirectional and unifactoral geological fixity in the world's oil bounty? Will the world's fate really be decided by restrictively defined oil reserve numbers already known and determined once and for all?

In the Peak Oil model, a definition of oil is chosen primarily for the purposes of measurement. It is placed in a specific framework. The chosen definition is then treated as if inviolate, the framework as unchangeable. It becomes a straitjacket into which all point-in-time estimates are squeezed. Projections of oil production are then made in order to calculate the numbers of years before the rate of oil output peaks. In the more sophisticated models, the estimated oil reserve stock is depleted by projecting annual consumption flows along a statistically estimated decline curve. From such models are deduced, first, a discovery peak, second, the production peak that is now said to be imminent, and third, the end of oil.

These models are usually based on so-called "regular" or "conventional" oil. As a result, they exclude various other kinds: oil from coal and shale; bitumen and derivatives; heavy and extra-heavy oil; deepwater oil from a depth of more than 500 metres; polar oil; condensates; natural gas liquids (NGLs); and oil from renewable resources (for example, ethanol derivatives, BTL). The definition is highly restrictive. Forecasts derive from estimates based on this definition, though they involve too – as we shall discuss – assumptions about such matters as oil resources, available reserves (proven, probable, possible), cumulative production to date, recovery rates and yet-to-find (YTF) or undiscovered oil.

The classic models built by Hubbert, Campbell and Laherrère are all based on fixed parameters.[50] Peak Oil theory adopts a very specific and limited view of geology and supplements it with assumptions that largely exclude the messy worlds of economics, technologies and/or geopolitics. The results are estimates of a *quantum fixe*.

This *quantum fixe* is one of the central planks of Peak Oil thinking. A second is the view that oil is exceptional, even unique. Campbell in *The Final Energy Crisis* advises us that oil is "not just another commodity obeying the normal laws of supply and demand as entrenched in economic theory and dogma".[51] The claim of the uniqueness of oil has over the years become a refrain, though the grounds for this claim are unclear. Is it because oil is the lifeblood of the global economy? Because it dominates traded energy and transport fuel? Or is it because oil has (supposedly) limited substitutability? All these arguments may be questioned.

It is difficult to see why any of these reasons should remove oil from the realm of economics. Quite the reverse, one might suggest. The global role of oil makes it subject to worldwide economic realities, including the geopolitical, that are often grounded in deeper economic and strategy interests. The very tradability of oil ensures that questions of oil markets are shot through with economic considerations. A similar point holds with regard to the use of oil as a fuel for transport: the choices made are bound up with economic questions concerning supply, price, substitution, comparative costs and commercial technologies. Peak Oil theory and its claim here is simply inconsistent with the way that the world oil industry operates.

Much is made of oil's difference from minerals as a reason for differentiation and the treatment of oil outside the realm of economics. Coal fields, it is argued, may cover a large area and be mined where the seams are thickest or most easily accessed. Mining decisions are subject to the normal economic laws. They depend on changes in costs and prices. But oil, it is claimed, is different – and not only because it is a liquid, concentrated by natural geological processes. Oil, it would appear, "is either present in profitable abundance, or it is not there at all, as is confirmed by a glance at the oil map, showing how oilfields are concentrated in clusters separated one from another by vast, barren tracts that lack the necessary geological conditions". Is this so?

Of course, the question of where and when oil reserves are exploited depends on choices based on economic criteria (investment committed, expected return, profit margin and a set of commercial and non-commercial risks). As companies experienced in pursuing marginal fields will attest, even the smaller fields have their day in the sun when economic conditions are right. Economics plays a central role in the definition, discovery and extraction of oil. That is how the industry works and always has worked.

Worldwide basin spread and the distribution of oilfields are not simply known once and for all. Knowledge of a basin's existence is not enough to know about all it might contain at this time. Basins have defined spatial areas but undisclosed prospectivity. Estimates are commonly adjusted in the light of advances in exploration, geological thinking and technology.[52] It is difficult to see how anyone who, like the author, has over many years viewed hundreds of basin and field maps, corporate asset and block positions, and state acreage licensing block overlays on regional geological prospects, and listened to thousands of presentations by leading geologists and senior management from the world's best companies, could believe that oil "is either present in profitable abundance, or … is not there at all". Questions of scale and distributive spread are inherent in petro-cartography and habitually require economic analysis.

The very term "profitable abundance" needs scrutinising even if it appears self-evident. What exactly is "abundance"? Is it really possible to define it in pure terms, independent of economics? Reserves that under one set of conditions appear to be marginal plays can, when conditions alter, look more abundant. This happens with regularity. Considerations of economics are, therefore, built into both halves of the phrase while profit implies cost and return – concepts fundamental within economics and economic geology.

Yet, despite these defects in theory and perceived reality, there is something seductive about the Peak Oil model. Its assumptions – that all is known and that everything is fixed – offer an appealing sense of certitude and implied stability. The prevalence of numbers in the model appeals to our modern, Cartesian attraction to measurement and quantities. The theory is simultaneously exhilarating – it is as if we have discovered the very calculus of worldly hydrocarbon goods created by the Gods – and humbling too: evidently our destiny has been written in the oil rocks, regardless of our will. As a bonus, the model's exclusion of above-ground factors – government policies, technology development and application, economic variables, corporate strategy – enables the Peak Oil travellers to bypass the untidy, unpredictable, socioeconomic world altogether.

Purists in Peak Oil claim that the numbers for stock reserves are already known and will not, except within very narrow limits, ever change. Among the specialists inside the world industry, however, there is much disagreement on such matters. Many geologists with contemporary operational experience – many of them engaged in the search for new plays in old basins or new ventures in frontier basins – consider that the world below ground is not at all well known. There is much unknown and a great deal undiscovered. That is one reason the number of exploration companies worldwide has proliferated.

Those working from the best world oilfield databases argue from different definitions and perspectives that Peak Oil estimates of reserves are wildly wrong (by anything up to a factor of two or even three). This conclusion, moreover, concerns only the data already known and the conditions that prevail today. Further allowance needs to be made for the possible effects of factors such as further worldwide exploration, enhanced basin experience, and wider access to resources and potential reserves. The Peak Oil model, we need to remember, is based on an aggregation of selected data from separate countries and fields – yet because many of these datasets are weak, so too are the resulting Peak Oil estimates. There are also many countries for which the data are weak and even some where it is not well known except within the state concerned.

Peak Oil claims that geology has in effect reached its limits. There is little or no room left for any further play of the geological imagination. Though some reserves doubtless remain to be discovered, their significance may be estimated accurately now, it is claimed – and this quantum turns out to be small. Similarly, the impact of oil held in potentially commercial reserves is also easily estimable, for it exists, apparently, in unchanging ratio to the whole. (This, it should be noted, is not the view of the SPE, which sees that ratio as dependent not only on economic variables but also on a gamut of changing and unknown conditions influenced by both Heaven and Earth.)[53]

This is not, however, to argue that the supply of oil is entirely elastic. There are those who argue that there exist, in effect, no practical reserve limits. They argue that there is no limit to the price of crude – and as the price rises, supply will simply expand as more reserves become profitable. This argument is the opposite of Peak Oil theory: it presents the question of oil supply as a purely economic one. But in practice, things are not as simple as that. We cannot be certain that cost-effective technology will be developed to enable the exploitation of all reserves. The future will depend too on government policy regarding such matters as allowable access, resource nationalism, taxation and contractual terms. It will also depend on upstream investment in exploration and development, and hence on corporate strategies including those of national oil companies.

Between the *quantum fixe* of Peak Oil theory and an infinitely elastic *quantum flux* is found a vast and uncharted middle ground of possibilities. Estimates of the point at which the supply of oil becomes problematic may vary by several decades. The further into the future that point lies, the greater the possibility that our dependence on conventional oil will lessen in competition with such energy sources as commercialised unconventional oils, natural gas and LNG, and in time diverse renewable fuels.

Given these criticisms of the Peak Oil model, we might ask how it is that its theorists generate such problematic conclusions in the first place. The answer lies, at least in part, in the methodology: in concepts and the numbers deployed. The problems begin with the selection of reserve data. Some of the data available from the USGS, the IEA and governments, as well as well-known and much-respected databases held by IHS, are inconsistent with Peak Oil's data and its findings. Such data tend, however, to be regarded by Peak Oil theorists as untrustworthy. In the nature of things, there are indeed difficulties with such data. All data are imperfect. Certainly they are subject to constant revision. But data need to be selected or rejected on the basis of objective criteria. Without these, the danger is that inconvenient statistics simply get wished away. In some measure, disputes over data veracity lie at the heart of Peak Oil's own credibility.

There are problems too with the assumptions made over both known and unknown data. A forever fixed resource/reserve ratio is not a concept found in the SPE formulation. Nor is it found in world oil history. Companies do not operate on this narrow fixity model either. They seek to expand the viable reserve base within their accessed resources. We have seen too, for example, that a relaxation in the assumption that the amount of YTF oil exists in constant ratio to current known reserves would generate larger future potential. All these considerations would have a direct impact on Peak Oil forecasts. The danger is that the *quantum fixe* model becomes tautological, the conclusions merely confirming what the model has been designed to show. After all, even the most refined models – those using algebraic formulations built on parabolic fractals, as in the case of Laherrère, for example – are only as good as their methodology and the numbers plugged into them.

We should remember that, for all the claims to immutability, frequent adjustments to Peak Oil models have been made, and have to be made. In particular, estimates of reserve stock may be increased, with the effect that the date at which oil production will peak may be pushed further into the future. This has occurred on many occasions already. Somehow, however, the date forecast for the Peak Oil moment is always just a few years ahead of us.[54] There is about all this a sense of teleology: the main features of the model – the *quantum fixe* of oil reserves and hence supply with the imminent peak of production – emerge, it seems, as unfalsifiable.

It is illuminating here to contrast Peak Oil theory's approach to such key questions as definition and estimation with that adopted by the Society of Petroleum Engineers. The SPE devotes enormous attention to its definitions, on which it publishes guidance in *Petroleum Resources Classification Systems and Definitions*.[55] These definitions are the most widely used in the industry. Their use is considered best practice.

The problem of how to standardise definitions has engaged the industry since the 1930s. Now, with assistance from the American Association of Petroleum Geologists (AAPG), a standard framework has been produced. Further work continues, especially in the light of recent problems concerning Shell's reserves.[56] While no doubt appearing arcane at first glance, the issues raised by such definitional questions have major implications for the Peak Oil debate.

The SPE's Schema of the Resource Classification System is given in Appendix 2. It provides a taxonomy comprising: reserves (proved; probable; and possible); contingent resources and prospective resources (both subdivided according to estimates); and unrecoverable resources. Estimates based on these definitions clearly require the application of integrity, skill and judgement. Such estimates are the product of multiple factors related to geological complexity, the stage of exploration or development, the degree of depletion of reservoirs and the availability of data. As a consequence, the estimate of oil initially in place (OIIP) at any given date "can be subject to significant uncertainty and will change with variations in commercial circumstances, technological developments and data availability".[57] At any time, a portion of classified unrecoverable resources "may become recoverable resources in the future as commercial circumstances change, technology developments occur, or additional data are acquired".[58] Fixity as determined now, once and for ever is not the philosophy behind such a schema.

Proved reserves are defined by the SPE as those that are "commercial under current economic conditions, while probable and possible reserves may be based on future economic conditions".[59] Factors influencing the determination of present and future proved reserves include current and prospective "operating methods" and "government regulations". According to the SPE, "economic conditions" include *inter alia* relevant historical prices and costs, contract obligations, corporate procedures and state regulations in reporting reserves. The economic and the uncertain pervade this set of standard definitions.

Unproved reserves may be influenced by the same factors. Similar considerations also affect the definition of probable reserves. Caveats, related to probability measured at 50% of likely oil recovery, include incremental reserves from potential infill drilling, reserves attributable to improved oil recovery, and so on. Possible reserves have only a 10% probability of recovery, but again are subject to economic factors as well as many as yet unknown considerations.

77

Clearly geological determinism plays no part in the SPE's definitions. Even estimates of current crude supply are subject to a range of economic criteria. This is even truer of future supply, given the potential for commercial, geopolitical and technology change. The SPE schema, therefore, calls into serious question the concepts on which Peak Oil theory is founded. It reveals that the notion of reserve measures is flexible. There is within the schema an evident dynamic potential: reserves may be reclassified over time. "Probable" and "possible" reserves may become "proven"; some "contingent resources" may in time come to be counted as "reserves". The schema thus allows for non-linear dynamics. There is also a case to be made that it requires such a view. Juxtaposed to this formulation, conceptual schemes confined to the linear (or even unifactoral) look seriously anachronistic and fatally flawed.

None of this means that Peak Oil theory must be written off *in toto*. It at least provides a pedagogical device or starting point for reflection. As such, it may help to shed some light on the dark world of oil, though its narrow lens ensures that such illumination is akin to torchlight. It does very little to illuminate the future, with its as yet unknown technologies, advancing geoscience, complex economic variables and to-be-discovered political dispensations.

8

Homo economicus: why oil is no different

In the 1970s and even into the late 1980s it was common to find oil companies employing chief economists. The chief economist was the person who produced the forecast for the price of crude. Much has changed since then. New tools – in the management of risk and uncertainty, for example – have made the price forecast less critical, and companies are less keen to publicise their economic thinking and internal criteria for investment. The upshot is that today far fewer oil companies employ chief economists. But at the same time, more people in oil companies employ economics by either design or default.

It would be a mistake, however, to conclude from this that economics has somehow exited the oil industry. Quite the reverse is true: economics permeates its thinking and practice. All key decisions are filtered through an economics matrix, whether they concern engineering, geology, risk mitigation, or parts of corporate business. This is especially true in the management of upstream strategy, which is where most critical decisions on matters such as reserve management, new ventures, country entries, oilfield developments and production choices are made.

The field of corporate strategy provides a good example of an area in which economics plays an essential role. Strategy management is crucial for survival and success in the oil industry. Curiously, it is a concept about which Peak Oil theorists have little to say: it is as if somehow "the facts" (the geological or reserves facts selected) are supposed to speak for themselves. In reality, the management of strategy calls all the time for judgements to be made. It is on strategic decision-making that many of the best minds in the industry's senior management are engaged – especially in the upstream business. It is in the field of strategy too that many of the greatest thinkers in economics (and the allied field of game theory) have excelled.[60] The art of strategy and the art of economics (most experts in the subject will confirm that economics is an art as much as anything else) have long been intimate bedfellows.

As we have seen already, Peak Oil theorists have little time for economics and economists. This is largely, of course, because the geological determinism of Peak Oil theory leaves no room for their ideas. However, a number of specific charges are made concerning their somewhat malign influence. We have already dealt with one such claim, namely that oil as a unique resource is not subject to the conventional laws of economics. Here we will consider the remaining charges.

Economists are often suspected by Peak Oil adherents of being in thrall to the open market. They are said to use neoclassical models that are incapable of handling the special role of energy resources in general and the unique role of oil in particular. Economic theory has been caricatured by its Peak Oil critics as a simple-minded belief that "supply and demand determine every-thing". This charge is compounded by the accusation that the focus of such theory is exclusively short-term (which would indeed be inappropriate for the oil industry). This reductionism is at odds, however, with the long history of debate within economics, during which the concept of the market as an "invisible hand" has been subject to detailed scrutiny, critique and refinement. Similarly, the debate has yielded a variety of theories of the long term. It would be a mistake to characterise the economics profession en masse as either having a blind faith in the workings of the free market or suffering from the myopia of short-term determinism. Indeed, economics is a rich tradition of thought and complex ideas spanning some 250 years of intellectual endeavour.

Much of the attention devoted to the oil market by econo-mists focuses on the issue of price. This is one of the areas in which disagreement with Peak Oil theory is most marked. The classic Peak Oil view is that the price of crude does not matter very much. No one active in the oil industry operates in such a fashion. In the case of conventional oil, they argue, most reserves would be profitable even with prices as low as $15–20 per barrel. Indeed, many Middle Eastern reserves would, so the argument goes, be profitable at under $5 per barrel. As a consequence, "not much is added [to oil supply] by increasing prices above say $20."[61] Curiously, however, price is granted some importance in the case of non-conventional oil. This view of virtually complete inelasticity for conventional oil conflicts with the implications of the SPE Schema (Appendix 2), in which the very definition of reserves is in effect responsive to price, among other variables. After all, companies' decisions on, say, whether it is worth while to run the risks involved in new frontier development will clearly depend on anticipated profitability – and hence costs and price.

A variation of Peak Oil's critique of so-called free-market economics (it is very much a straw man in Peak Oil's hands) comes with the observation that the oil industry is not, in any case, perfectly competitive. Increasingly, the critics argue, oil production is concentrated in the hands of a few major players. It may be granted here that it would certainly be a mistake to seek to apply models of perfect competition to an imperfect market. That, however, is not necessary – for the simple reason that, in the course of its evolution, economic theory has (particularly since the 1930s) developed models for understanding and predicting the behaviour of oligopolies, oligopsonies, cartels and monopolies. Nor is it necessary to discount all economics when the principles it confirms can be applied widely.

Economists are also often accused of ignoring resource constraints. Well-known economists on oil such as Harold Hotelling, Adelman, Odell and Lynch have apparently got it all wrong.[62] In particular, they are said to be in denial over their handling of the concept of "ultimate recovery" in oil. Crucially, argue Peak Oil commentators, this is a finite resource that economists insist on treating as if it were infinite, thus ignoring the process of depletion. This is just not so and at best reflects misunderstanding.

Many of the accused have discussed at length the related

matters of resource constraints, ultimate recovery and depletion. This is particularly true of Odell, with his 40 years of experience in the business and a 267-page book devoted to just such issues. In practice, in-house economic evaluations by companies typically treat reserve quantum as a finite sum initially, though one that may be treated over time in dynamic terms as potentially subject to a wide array of variables – for example, discovery or reserve growth beyond initial new field wildcats (NFW), potential deeper oil pay zones, the role of associated liquids or hydrocarbons, step-out reserve accumulations or EOR potential.

It is not only theoretical economists who grapple with such concepts. Practising economists in the upstream business engage with the same concerns, particularly in the management of assets and risk. Campbell, however, considers that the risks that they attempt to assess and manage – risks concerning exploration, tax, contracts, politics, terrorism, the environment, corporate status, and so on – are "greatly exaggerated".[63] Such wholesale discounting of multiple risks would be greeted with amazement by players in the oil industry, where virtually all management decisions have risk profiles embedded within them. Such economists are not in denial: they simply think differently from Peak Oil theorists. Indeed, about the real oil world it is Peak Oil that is in a state of denial.

In a parallel criticism, economists are often said to use the concept of the reserve/production (R/P) ratio uncritically, failing to recognise that one day production may fall close to zero. However, seasoned economists recognise that R/P ratios have a limited utility and require careful handling. In particular, a

THE BATTLE FOR BARRELS

distinction needs to be made between the use of R/P ratios for a field and their use in estimates for an entire country or even globally. Aggregation reduces the robustness of the statistics. At the macro-level there does of course need to be some notion of the relationship between reserve presence and the rate of production. For this purpose we need the concept of a "dynamic R/P" applied and adjusted for both variables. Because of course future conditions are unknown. A proxy for any future resource scarcity is often subsumed into the crude price as a heuristic device. This practice contradicts the criticism that reserve depletion is simply ignored.

The Peak Oil critique of economics concerns not only matters of theory, but also those of practice. The problem with economists, it is claimed, is that they just "never get it right". They lack "skill in predicting and understanding events". They make "false economic evaluation[s]" of prospects (though who would do better is not made clear.) As a result they are simply "failing" the companies that employ their services.[64] The criticism of practice focuses particularly on the upstream business. Here economists are accused of distorting the oil search by, in particular, sustaining frontier exploration "for many years longer than justified by the geology". None of this can be credited with correspondence to corporate reality in which managements typically arbitrate the technical and non-technical judgements of geologists, engineers and economists to make decisions on strategy and investments. Yet economics, it seems, corrupts the pure world of geology. If the oil world were made of ice cream, geology would be the only vanilla.

It is, however, difficult to see why frontier exploration is viewed so negatively. There has always been a frontier. There probably always will be. A frontier is, after all, not simply some now known or limited geological zone: it may be a new country or basin, a different environment, technology not yet applied, some innovative financing structure, a changed geopolitical context, or some mix of all of these. The history of the industry is one of constantly pushing back these frontiers, with or without the assistance of economists but almost always with some dependence, explicit or implied, on the discipline of economics or ideas drawn from it.

In evaluating the contribution of economics to the oil industry, it is important to do more than simply rebut the Peak Oil critique. That would be facile and give far too negative a view. We need also to set out the positive contribution of the discipline. From an initial focus on factors of production (including natural resource constraints) and an early preoccupation with diminishing returns (the original depletion theory), through theories of oligopoly, cartels and monopoly (including complex bilateral monopoly models), to the emergence of theories of economic growth, of the firm, of corporate structure and the rich literature of over half a century of development economics, the traditions of economics have enriched our theoretical and empirical understanding of both the world in general and the oil world in particular. The roll call of world-class thinkers in economics is a long one.[65] Their ideas, as Maynard Keynes once noted, "are more powerful than commonly understood".[66]

Out of this development has come a variety of views informing all societies on matters as varied as oil wealth, oil poverty and growth; competition and technical change in the hydrocarbon industry; the role of time in energy investment; petroleum input–output productivity; economies of scale in oil development; externalities in corporate upstream portfolio management; petroleum business cycles; oil production functions; upstream technologies; oil markets and substitution; crude market determinants; great power oil strategy and game theory; public resource goods and the oil state; value and valuation of oil assets; oilfield investment and cost-benefit analysis; oil and debt; upstream portfolio investment; knowledge and applied reserve growth; disequilibrium in petrodollar balances; non-linear theory in upstream dynamics; and, of course, oil supply and oil demand. These are a few of the ideas (often combined with notions of scarcity, resources and reserves) that have been tested in economic theory and in empirical analysis in the industry as well as in many professional economics and energy economics journals.[67]

The Republic of Economics, as David Warsh has noted, is a "big place".[68] It is replete with factions and periodic civil wars in the contest for ideas (a perusal of the naming of Nobel Prize winners in Economics is enough to suggest why). As a result of this endeavour, the world portrayed by such economists appears richly detailed, intensively fluid, multi-organic, ultra-complex, and maybe even occasionally vague – which is how the modern oil world is found.

Perhaps the greatest contribution of economics to the oil industry lies in the treatment of the complex matters concerning time. Economics has made countless inputs into matters both qualitative and quantitative in the upstream business and in the oil markets of the world. These include, for example, issues concerning pricing for crude oil qualities, oil futures markets, commercialised associated gas and condensates in an oil-gas field.[69] The use of applied economics in discounted cash flow analysis for oilfields and prospects, the valuation of oil reserves using net present values and in capital asset model realisation, is no secret brew. All these activities flow from the economic theory of time and the related concepts of monetary value, capital value and its derivatives.[70]

No apologia for oil economics would be complete without noting one particularly rich irony. For all its contempt for the house of economic theory, Peak Oil theory rather closely resembles one particular room within that house – namely Malthusian theory. Juxtaposed to Peak Oil theory, the Malthusian scenario of dramatic crisis caused by demand outstripping a limited resource environment looks uncannily familiar. There are, however, differences between the models. Malthus's theory included some self-correcting, almost Darwinian, mechanisms. Much Peak Oil theory has even more limited "dynamics", though several post-peak theorists foresee in the *fin du monde pétrolier* a huge demographic catastrophe. According to them, we will reach a new steady state only with a diminished world at very much lower standards of living.

How does Peak Oil theory look, once the supposed debunking of economics has itself been deconstructed? Now it can be seen that the elegance of its statistically adjusted depletion curves and the comfort of its supposed mathematical precision are made possible only by its wishing away of the messiness of the oil world, with all its state and corporate strategy manoeuvres, economic decision-making, financial calculation and realpolitik.

Such lack of realism may be a product of intellectual fashion, and it is a fashion that has had some history. In this arena Peak Oil's gurus are distinctly arriviste. It is possible to trace back a long tradition of alarmism. During his term in office in the 1970s, President Jimmy Carter had come to the conclusion that all proven reserves oil would be used up by the end of the next decade. In 1920, the USGS had put world oil reserves at a mere 20 BBLS. In the early 20th century there were regular predictions of oil famine: in 1914, the US Bureau of Mines said that America would run out of oil in ten years. And back in 1855 the rock oil found in Pennsylvania had been predicted to disappear, the victim of reserve depletion. The peaks have been shifting for a long time now.

The gloomy, pessimistic 1970s, in particular, proved fertile soil for the growth of doomsday scenarios, such as the Club of Rome and the Limits to Growth model, impending population bombs and the collapse of capitalism. The publication in 2005 of *Limits to Growth: The 30-Year Update* is a reminder that our era is in some ways comparable. It is unsurprising that in our troubled and uncertain times a theory such as Peak Oil should flourish. Angst feeds on angst.

Some Peak Oil conditions may, of course, happen eventually to coincide with short-term or long-term empirical events. We may indeed witness, for example, times when the supply of crude oil is exceptionally tight and the secular trend of prices leads sharply upwards. It may be that today's bottlenecks in oil infrastructure will require a long investment cycle to widen sufficiently, specially in the areas now closed, sanctioned or at war. That, though, despite what Peak Oil theory may suggest, is hardly geologically determined. Such problems are not written entirely in the rocks.

9

Peak Oil theory and the evidence

We have already seen that one of the unsatisfactory aspects of Peak Oil forecasts is the definitions used, which are highly restrictive. Clearly, the more the definition of oil is restricted, the more limited its supply appears and the easier it is to support forecasts of imminent peaking in production. These definitions do not in general correspond with the way the oil world works. Though oil traders categorise crude by quality type and grade, the consumer certainly does not: no motorists at the pump specify that they want fuel derived only from regular conventional oil. Some companies' investment strategies now target certain types of oil (heavy oil or GTL, for example). Many large players are found in the deepwater oil world. In general, however, Peak Oil's method of definition does not correspond to the way that corporate oil reports its reserves, generates its upstream asset portfolios, or makes new ventures decisions. Crude oil is an entirely fungible commodity mainly categorised in the industry according to economic criteria (principally cost and price). In the end, Peak Oil theories usually have to admit at some point that it is not possible to draw firm lines between regular conventional oil and other types.

Peak Oil theory tends to operate at a high level of abstraction. Its forecasts centre on extrapolations from estimates of reserves and derivative global oil supply. These estimates are based on the aggregation of estimates made at national or regional and country level. Though comments are made about particular fields or types of field, Peak Oil theory has never produced a field-by-field supply forecast. The estimates it does use appear to conflict with the Petroconsultants database owned by IHS. There are glaring discrepancies between Peak Oil estimates of reserves and IHS's more expansive view.

A further weakness of the Peak Oil model is the engineering paradigm on which it is built. So much is excluded by this choice of paradigm that the model barely floats on the choppy seas of reality. In the world of engineering, there are usually reliable laws and measurements on which to base calculations of, for example, size, weight, volume, scale and stress. Peak Oil theory presents itself as a sort of physics of oil reservoirs. In doing so it in effect looks at the oil world as through a microscope (a device designed to exclude everything except the specific material one wishes to enlarge).

The static nature of the Peak Oil model is another fundamental problem. However accurate the estimates and the equations for relationships between variables at one particular moment in time, the model is liable to go haywire as soon as questions of dynamics (time and, with it, complex interdependencies between variables) are introduced. Peak Oil's static model is thus a prisoner of its own formulaic calculus. The real oil world does insist on changing all the time, often in unpredictable ways.

It is not surprising that the result of these imperfections in Peak Oil theory is a series of constantly shifting and contrasting forecasts. Running after fast-moving realities, Peak Oil thinkers have announced "adjustments" based on new data and refinements to their models. The result is a Dante-esque shifting of peaks, some examples of which we have already noted, so let us examine the theories in a little more detail. They reflect concerns at the heart of the predictive dilemma within Peak Oil theory.

Deffeyes argued that combining Hubbert's 1982 estimate of world oil reserves (between 1.8 TBLS and 2.1 TBLS) with his 1956 method yielded a forecast for a peak around 2001–03.[71] In *Hubbert's Peak* (2001), Deffeyes himself seems to settle for 2004–08, though he noted that "the event may even occur before 2004" (however, with "a lot of unexpectedly good news" it could be postponed to 2010). It was also noted that using a generous reserve estimate of 2.1 TBLS with a Gaussian curve "fit" produced a date of 2009. In the end, *Hubbert's Peak* settles on a date of 2005 (give or take a year). In the late 1990s, other analysts using Hubbert's model forecast that oil production would peak in 2004–08. The analysts included Campbell and Laherrère (*Scientific American*), C.B. Hatfield (*Nature*) and R.A. Kerr (*Science*).[72] Campbell has now revised his forecast towards 2005–10; Laherrère now forecasts the event somewhere within 2010–20. Times change and so does Peak Oil.

To these forecasts should be added those of the nouveau peakists who, it would appear, sift the forecasts of others rather than do the hard graft of forecasting themselves. These form a wave of second-order peakists. Bakhtiari opts for 2006–07; Simmons is reported to favour 2007-09; Skrebowski cites sometime after 2007; David Goodstein expects the peak before 2010. Most Peak Oil commentators now seem to favour 2010. A consensus around that year seems to be emerging within ASPO.[73]

Differences outside ASPO have also been quite marked, though in general forecasts produced by economists have been more sanguine. The World Energy Council thinks that Peak Oil will arrive "after 2010"; PFC Energy has cited 2015–20; the EIA reference case mentions 2016, though in 2000 it cited the range 2021–67, with 2037 most likely; CERA has suggested sometime after 2020 and perhaps a plateau; Shell says it will be after 2025; Odell has suggested 2030 (with a seamless conversion to non-conventional fuel, so that a real oil-liquids peak occurs only towards 2060). Lynch does not seem to see any definitive peak and at least postulates that this is not the germane problem.[74] The Peak Oil game is not one that all need to play. In *Trendlines* Freddy Hutter considers a number of estimates varying by as much as 45 years (that is, 2010–55) with peak production ranging from 85 MMBOPD to 120 MMBOPD (that is, 40% above current levels).[75] There is a range of forecasts too for the date at which production slips below 10 MBOPD, although none of the forecasters are likely be around to know whether they were right: their dates vary from 2070 to 2140. What is noticeable is that when all is considered, a wide range of dates can be found.

The great advantage for Peak Oil theory of constantly shifting the forecast forward is, of course, that it increases the chances of eventually getting something right. It is embarrassing to note, however, that several of the dates that have been forecast for peak production by eminent sources have already passed. In 1972 the UN set the date at 2000; Hubbert (variously from 1974 onwards) forecast 1991–2000; in 1972 BP's chief geologist, H.R. Warman, foresaw a peak before 2000; in the same year Shell's chief geologist, A. Hols, agreed with Warman; in 1977 Ehrlich pitched for 2000, as did the World Bank in 1981. As Odell wryly notes: "One can argue for a world which over the last 30 years of the 20th century since 1971 was 'running into oil' rather 'out of it' as so widely forecast in the 1970s."[76] Now everyone seems to feel the need to produce a forecast – perhaps fashion or political correctness requires it. One thing, however, is certain about the current competition to become the new Nostradamus: as volume rises, past exactitude decreases.

There is considerable competition between forecasters, many of whom claim to speak with certitude. For some, forecasting is virtually a business. The divergences are only partly explicable in terms of definitional differences. Various fault-lines have emerged between forecasters – between the technical and the non-technical, for example – and between economists and those who adopt a purely geological perspective. Forecasts also vary according to commercial and social concerns and political or ideological alignment. Those working inside corporate oil (managers of long-term strategy, for example) might be expected to be the most concerned about Peak Oil – they, after all, have the most to lose. However, they tend to regard the debate as unrealistic and overly academic, concentrating instead on more practical matters close at hand.

There is less dissent over the direction of recent geopolitical developments but much disagreement over its interpretation. The scenarios of benign globalisation that were popular in the early 1990s have given way to a sense of fractured globalisation, in places bordering on a kind of structured anarchy. It is undoubtedly the case that oil is now sourced from a range of imperfect, unstable or unsavoury regimes and others that still invoke concern in the West. Neither West Africa (notably Nigeria's delta region) nor the Caspian states exhibit the benign conditions of pure market theory. Few with operational experience would treat them in such a manner.

There is also worldwide terrorism, and the global oil supply chain has become more vulnerable: witness Saudi Arabia, Iraq and elsewhere. Russia's strategic control of oil assets has been reasserted and is concentrated in the hands of Rosneft and Gazprom, probably at some cost to potential future output and certainly in regard to inward foreign oil investment. Iran threatens to unleash an oil crisis, potentially in the Gulf of Hormuz, if its nuclear ambitions are thwarted. It has the capacity to do so. Iraqi oil production has struggled to meet pre-conflict estimates of its full potential and this situation may prevail for some time. Venezuela has crafted a new oil diplomacy built on the "national oil control agenda", but this has compromised PDVSA's production trajectory. China and India have been more active overseas in securing equity oil and have acquired assets worldwide in an aggressive commercial manner. The Chinese initiative is viewed by some as a pre-emption on western oil security. Asian oil demand growth has an impact on the world's additional marginal production (perhaps at 50% or more for the coming decade or so). The US speaks about reducing its addiction to foreign oil and achieving energy independence. The oil market has undoubtedly become tighter, perhaps more so than for three decades. And Al-Qaeda has an oil agenda and an apparent strategy to accomplish it. All these geopolitical and strategy issues affect corporate oil's access to reserves and hence have some uncertain bearing on supply.

Many commentators claim that major geopolitical developments are "all about oil". Certainly geopolitics has persistently played a role in world oil, as the world oil industry's historiography, best explained by Daniel Yergin, illustrates.[77] It does not follow, however, that it is imminent depletion that causes all these developments, though some have offered grounds for thinking so.[78] The author commonly encountered such views at the 2006 ASPO conference: one leading light even claimed that the Lebanon–Israel crisis that year was "simply caused" by oil depletion. In Pisa, developments that would greatly expand the world's net productive capacity – new field developments (signalled in CERA's report and with much investment already in place) and developments concerning deepwater oil and Canadian oil sands – were significantly discounted.

Thus even in world historiography, contemporary as it may be, there are divergent views on interpretation and causality. The Peak Oil *weltanschauung* connects all such events to a single causal factor, thus offering a "theory of everything" that allows no room for competing theories and interpretations (the Incompetence Theory of History, or Murphy's Law). Non-oil dimensions are dismissed as epiphenomenal, marginal and even delusional. It is as if the world is engaged in a vast dance born of collective oil conspiracy, operating perhaps at an unconscious level. Well, it is one thing to suggest different theory or ideas. But what if the numbers on which all this is based are wrong?

I O

Counting crude oil beans

In Peak Oil's mantra much hinges on estimates of oil reserves and capacity, yet how to make sense of such estimates is far from self-evident. Not only are there many sources, not all of them public; there have also been charges of incompetence, manipulation and secrecy. About some of these – especially those regarding Saudi Arabia – Da Vinci (or Dan T. Brown) would be proud. We could all do with an oil codex.

To arrive at judgement in estimates, there is, evidently, a need for all manner of operations: decoding, intuiting, adjustment, revision, backdating, plugging of gaps, redefinition, discounting, number crunching, guesstimation, speculation, and all manner of "fixing". It is strange, in view of the contentiousness of such inputs, that any forecasters could claim their outputs to be incontestable or inviolate. Differences between data inputs are an important cause of variance between forecasts. Which data are selected depends on many complex factors, including historical chance, technical expertise, corporate status, alleged vested interest and invested credibility. Yet reputations are staked; good guy–bad guy syndromes emerge; and the watchword becomes,

"My data are best, yours are poor, concealed or manipulated or even corrupted."

The data predicament lends itself to intricate conspiracy theory. Well-known data sources, especially the USGS, are dismissed as "highly questionable". There has even been comment in jest that data merchants act like "spy companies", though it is hard to reconcile such images with any known organisation's modus operandi and overt public character: nobody familiar with them would liken them to secret and slick CIA, KGB or Mossad-style operations.[80] But oil, Peak Oil commentators assure us, is not just a commodity and so nefarious interests must therefore be at work. The spooks are everywhere.

The *fons et origo* of all this disputed material are the prime data merchants. The most authoritative, according to Peak Oil thinkers and others, is the IHS world oil and gas reserve database built on a field-by-field basis. Other sources include the Wood Mackenzie database, proprietary databases held in some OECD countries (several now within the IHS stable) and annual data published in industry "bibles" such as *Oil & Gas Journal* (*OGJ*), BP's *Statistical Review of World Energy*, and *World Oil*. To this list can be added many national databases found in government ministries and licensing agencies (Russian data records are thought to be prolific) and Securities and Exchange Commission records on companies from filings often compiled and reported upon by entities such as J.S. Herold and Petrocompanies (now Evaluate Energy). There are also many corporate portfolio data and internal databases held privately by public and private companies around the world. No single individual or group holds all these databases, which if combined would constitute a vast resource. As a result, even what is known on world oil reserves is in some measure relatively unknown to many.

By no means all the world, of course, has been fully mapped, explored and appraised in geological terms. We live in a world yet to be fully discovered. There is, therefore, an unavoidable incompleteness about world data. Though existing databases may constitute reasonable starting points, one thing is for sure: no one enjoys access to all the data produced and no one knows all the unknowns. Livingstone and Stanley did not know everything about interior Africa before they began their safaris of discovery. It took many wanderings and false starts, along with decades of subsequent exploration, to sketch in even crude maps of Africa. Similarly with oil: we need to avoid claiming that all is known and acknowledge our areas of ignorance. It was a miscalculation for the 18th and 19th century "discoverers" of Africa to think otherwise – one that even Leo Africanus could have advised them of 300 years before their many fatal journeys.[81]

Let us consider what is widely accepted, by Peak Oil theorists and others, as a major database, namely the IHS database developed originally by Petroconsultants. It originated in its earliest format from the company's Foreign Scouting Service. Jean-Marc Lador, who with his colleagues guided the beast into the world, has described its origins best.[82] It has taken decades to develop the database, even to its current, incomplete, state. Over the years many editors and technicians have grappled with problems of data integrity, verification and consistency that have become legendary. They drew on basin studies, bought data from

Elf and others, called on consultants and hired talented geologists. In 1974 LEXIS, the first computerised version, was born. Records were compiled in Dublin by young Irish geologists and geography trainees hired from 1975 onwards. Without staff sitting atop each of the then 4,000 or so fields covered, there was inevitably a good deal of necessary estimation.

Though their Dublin office was closed in 1987, some of its staff continued to work on the database. In the early 1990s LEXIS was transformed into an expanded database named IRIS21. For all the arduous labour devoted to refining the estimates, imperfections inevitably remained. Those working in the company's Economics Division often found oil reserve estimates at variance with what operators reported or reliable senior industry parties advised. Nonetheless, a basin database was eventually constructed. IHS acquired this database when it bought Petroconsultants in 1998. At this stage there were changes in staff and methods, and some Peak Oil commentators claim that much expertise was lost. Nevertheless, the process of amendment and refinement has continued. The most recent IHS audit threw up more reserves found than were ever recorded before.[83]

It is on this database – long-established, laboriously constructed, gradually evolving, much respected, but inevitably imperfect – that Campbell and Laherrère built their earliest seminal Peak Oil models. Logically, then, revisions to the database will require changes to the Peak Oil forecasts, though they will sit awkwardly with continued claims to certitude based on their early and continuing notions of a *quantum fixe*. Even allowing for dubious Peak Oil practices (creating "backdated" reserves), and even after relying on 2-P reserves (that is, at 50% mean probability), recent Peak Oil extrapolations will need to be continually revised (and who knows how often?). Even then, such forecasts will require interpretation in the light of, in particular, the elongation of the corporate upstream corporate chain caused by the plethora of small companies born over the last decade or so. This is an area that is weak in Peak Oil's calculus. State players have also become much more important, though Peak Oil typically lacks access to their national and corporate data.

The significance of the IHS database and the issues around it are maximised by Peak Oil theorists' disdain for other industry sources. They consider *OGJ* data derived from questionnaires as suspect because it is said that they draw on "political" sources (that is, governments). BP's publications may apparently be dismissed because, up until 2003, it just reproduced the errors of the *OGJ*. Both then publish what is considered, apparently, "misleading information". *World Oil* is considered only marginally better. USGS data are considered particularly suspect – this otherwise much respected agency is regarded as a tool of the US govern-ment, combining definitional problems with an exaggerated view of the scope for new discovery and reserve growth. Estimates for the reserves of Middle Eastern OPEC members are also mistrusted having been, according to Peak Oil mythology, irra-tionally inflated in 1987–88 as a result of an internecine scramble for OPEC quotas. In the assumed malfeasance of governments in the oil data world, Saudi Arabia is given pride of place. It is said to have been "exposed" by Simmons for its secrecy, non-disclosure, dodgy claims to spare capacity and "unproven" or non-audited reserve numbers.[84] In effect the data world is a labyrinthine one that can be penetrated with skill but not absolute certitude. Here Peak Oil models select but one view, a highly restrictive one, for their reserve numbers.

At times the debates over data sources and the shenanigans that they are said to involve seem to bring us close to the realm of fiction. Or perhaps "faction" is the better term. Here the Geneva-based Petrologistics is characterised (more properly caricatured) by some as an operation with one employee and an office over a grocery store claiming to have a team of 20–30 harbour spies observing international crude/product cargo liftings. One could be forgiven for believing that the owner, Conrad Gerber, spends his time glancing at third-party estimates and either issuing supposed consensus reports to the media or plucking figures out of thin air. As someone with over two decades of knowledge of Petrologistics, its specialised product, and some of the sourcing and modus operandi, the author can summarily dismiss such images as ill-informed fantasy.[85] Petrologistics' modest headquarters are on a floor of an office block above Migros, a large retail chain in Switzerland. The company's data is extremely well organised, meticulous in its presentation of the numeric, and much respected by its blue-chip clientele. Its detailed data matrix covers a whole gamut of oil data specifics that no one could dream up, and no third party has such definitive sourcing at its disposal anyhow.

Peak Oil theory's need for certitude does raise the temperature of disputes over data. It is almost as if the gurus feel that nobody will give them the toys that they need to play their oil game. The result is a doctrinal war of words arising ultimately out of claims concerning the exclusive ownership of historical understanding, the authority of science and claimed authenticity in ultimate knowledge on world oil.

11

From Hubbert's Peak to Dante's Inferno

The voluminous writings now available on Peak Oil would probably overwhelm anyone wishing just to scan – let alone read – all of it. Fortunately, the intention here is more limited, namely to analyse the seminal texts. In particular we shall focus on the thinking of Hubbert – as refracted and developed by Deffeyes – Campbell and Laherrère.

Kenneth Deffeyes published *Hubbert's Peak*, a work of some erudition, in 2001. That is an important year in Peak Oil historiography, since it was also supposed to be when, according to the influential Hubbertian forecast, oil production was to peak. It didn't happen. But Deffeyes, far from abandoning the model, decided – as we have seen – to revise the forecast, initially by only a few years. An analysis of Deffeyes's response will do much to demonstrate the working of Peak Oil theory and praxis.

Deffeyes gave several reasons for his decision. He argued that new oil technology would have little future potential: the "wheels have already been invented". Deeper drilling outside the narrowly defined oil window was considered fruitless. The last remaining petroleum province was apparently the South China Sea – hardly a new Middle East and, in any case, bedevilled by unsettled boundary claims yet of much interest to China and other contesting parties.[86] Accelerated exploration takes at least ten years to yield fruit and so even new projects begun in 2001 would be unable to prevent production peaking within the decade. The case appeared to be a slam dunk.

Let us look at what has happened in the light of Deffeyes's arguments. We may note that a number of exploration cycle vintages were already in play by 2001. Indeed, the exploration game has been a vibrant one. Multitudes of geologists continue to scour the earth and companies have before and since taken acreage and established new ventures in many regions, not just around but even within the South China Sea. Exploration plays worldwide have been built not only on the basis of trendology within established places but also from technical strategies applied to upstream activities within new and old zones, and frontiers in Africa, Asia, Latin America, the Middle East, Russia and elsewhere. Deffeyes reminded us of Hubbert's rule: "If even one thing goes wrong, you get a dry hole."[87] This has happened often and, though true, perhaps should be overridden by Machiavelli's rule: "You can't know everything."

A major development in the contemporary oil industry has been the extensive activity of old and new independent companies. Their strategies depend crucially on their ability to integrate geological, commercial and economic insights with deal flow. They develop technology to lower drilling costs, enabling them to reach new targets or exploit previously non-viable reservoirs.

Deffeyes, in *Hubbert's Peak*, is aware of such strategies, yet his diagnosis fails somehow to recognise their continuing worldwide import. It is significant that Deffeyes's claim that economists such as Adelman and Lynch "state that oil discoveries depend on the level of investment put into the search" is only partially correct.[88] This is so and is only part of a more complex new ventures and discovery story.

It seems, however, that Deffeyes also underestimates the potential of secondary and tertiary recovery strategies (such as steam flooding and miscible flooding), which are all directly affected by the scourge of economics. This may be because of a focus only on the physical: the size and discoverability of oilfields. In effect, he retreats to an application of Zipf's Law on natural field distribution, while admitting that this is problematic in the case of smaller fields.[89] In a flash of contrarian insight, however, Deffeyes does wonder whether the "the entire Middle East is loaded with small fields".[90] This intuitive view is telling and so perhaps, we should add, this is potentially the case in many other places. Deffeyes also notes that undiscovered oil in the Middle East "is very likely the largest untapped supply in the world" – a view not always easy to reconcile with other Peak Oil texts.

While Deffeyes mounts a spirited defence of Hubbert, the latter's original concern, it will be remembered, was with oil production in the US. He argues that it is not necessarily fair to seek to disprove Hubbert's theory simply by shifting energy categories within the US or considering oil elsewhere. That point in itself does not, however, establish the original scientific credentials for Peak Oil theory. For all the elegance of the various fitted curves (Gaussian, Lorentzian and logistic, adding even parabolic fractals and creaming curves), one has to ask on what basis they can be used to predict the future.

As pointed out, the Hubbert model had been applied, with modifications, by Campbell to forecast a production peak in 2003. We have already seen that shifting dates are a feature of Peak Oil theory. Here let us focus on the key numbers that entail such shifts. Deffeyes wonders whether an estimate of 1.8 TBLS is too erroneous. He shows that an estimate of 2.1 TBLS would yield a peak forecast of 2009. This, we should note, represents a slippage of six years. Bartlett notes that a doubling of the estimated ultimate recovery (EUR) inside the Hubbert model moves the peak date back some 26 years.[91] J.D. Edwards has sourced 14 estimates of EUR ranging from a low 1.65 TBLS to a much higher 3.2 TBLS.[92] The numbers matter. Bartlett's doubling means that the forecast peak recedes by an entire generation. Much can change within such a long period of history.

Deffeyes then applies variants of Hubbert's model, based on cumulative production and discovery. Again, he heavily discounts reserve estimates for the Middle East, a feature found within Peak Oil's litany. As a result he produces forecasts varying across the 2003–09 period, adding that he "honestly do[es] not have an opinion as to the exact date".[93] This is perhaps a wise position to hold.

One wonders, in view of the varying estimates, why a figure of around 2.0 TBLS has emerged as a sort of magic number for Peak Oil. For Deffeyes it seems to represent a kind of received wisdom. Hubbert, among others, used it, as did Bartlett in *Mathematical*

Geology: "it was a nice even number" (certainly rounder than the USGS's estimate of 3.102 TBLS).[94] Deffeyes's final view, we should note, is that: "There is nothing plausible that could postpone the peak until 2009."[95] Again, time will tell.

As we saw earlier, the leading contemporary Peak Oil advocate is Dr Colin Campbell. He has metamorphosed from company geologist to consultant to single-issue intellectual to his current status as an activist, impassioned writer, lobbyist of parliaments, writer of public letters to CEOs and sometime political commentator. Combining an interest in the technical and the political, Campbell has become the figurehead of the movement. He chairs ASPO's annual conferences, makes appearances in the media, provides web content and writes prolifically. *Oil Crisis*, which describes something of his long personal journey, is a work of immense industry. Its 397 dense pages include 44 pages of bibliography, indicating wide sourcing, and references to 118 self-written texts.

As we have seen, Campbell has often had to revise his forecasts. Along with his revision has come a shift, evident in the transition between *The Coming Oil Crisis* in 1997 and *Oil Crisis* in 2005. Much technical content is common to both books, but in the second there is a greater amplification of historiography, world affairs, demography, issues concerning food and renewable energy, and other matters on which geologists are not necessarily considered the logical source of expertise. Campbell's basic model, however, remains much the same as in the early works. Regular or conventional oil remains central (though in later works there is attention too to unconventional oil). In the world oil tank, so to speak, we have both discovered oil (that produced already, plus reserves) and (not much) YTF, the undiscovered oil. As we use this oil, so the tank steadily drains dry.

Campbell's global estimates for production and reserves, aggregated from country estimates, are taken from the public record (if not from the Petroconsultants' database). Perhaps earlier versions were based on this database and later adjustments made in the light of the public record: we do not know. The totals are "adjusted" to remove unconventional oil and correct "implausible numbers" (especially the "anomalies reported" concerning Middle Eastern reserves). These adjusted values are "multiplied by a factor": exactly what this is remains very difficult to discover, as elsewhere it is cited as a certain assumed recovery factor. The result is an estimate of to-be-used assessed reserves at "proved and probable" levels (depicted as not just recoverable reserves but those that will be produced) on which future production estimates are then based. From these calculations emerges an estimate, based on "experience and judgement", of the ultimate recoverable reserves (URR). Here 50% usage of URR indicates Peak Oil mid-depletion. After this mid-point, future production is calculated for post-peak countries at the current depletion rate. The situation for pre-peak countries is "judged" according to local circumstances and a depletion rate of 0–5% is applied. Some undefined modifications related to drilling, giant fields and discovery trends are then made. What is clear here is that some judgements are involved.

Within the model, therefore, much turns on choices made at various stages concerning the selection and handling of data. The outcomes are sensitive to individual selection and judgement – a gloomy set of judgements will produce a pessimistic forecast. Adjudications made about reserves in the Middle East reduce reserve estimates by a significant 300 BBLS, the reasons relating to the so-called "fraudulent" or unjustified OPEC reclassifications made during the quota wars. A very conservative view is taken of Africa, much at odds with contemporary industry views and experience (see Chapter 18). Higher recovery rates could be applied, but according to this model, based resolutely on the notion of a *quantum fixe* so at odds with the implications of the SPE's classification, this would merely deplete the world's stock of oil even faster. The tank would just drain quicker. What appears on the surface to be mere hard fact depends in any final analysis on a good deal of subjectivity.

It is instructive to track the changes over time from the model's inception in 1989. The original URR estimate was 1,650 BBLS. Now the figure usually (though not always) used is 1,900 BBLS (approximately 15% higher). The main changes appear to have been a revision upwards in 1989–96 to 1,750 BBLS. In 2001 an estimate of 1,900 BBLS was used (to include condensates, perhaps). The date originally forecast for the oil peak was 1999. By 2001 the forecast was suggested as likely to be a plateau over 2000–10. In 2004 it was for the next year and in 2005, curiously, for the year before. Now many say the date will be around or towards 2010. As the author had noted in 1996 with some prescience, predicting the future is a dubious undertaking – especially in an uncertain oil world. That has certainly proved true.

Let us consider the figures for future exploration potential.

On the latest count (ASPO 2006) it seems that reserves of only 790 BBLS may remain, though in *Oil Crisis* in 2005 Campbell estimated 906 BBLS (but then URR was given, inexplicably, as 1,850 BBLS, so perhaps condensates were stripped out). The current estimate for YTF is low (142 BBLS, of which 28 BBLS are assigned to "unforeseen discoveries in new or currently producing countries"); yet as we were told (even in 1996), the world has allegedly been explored sufficiently to discover all major provinces. On any view, however, not all major provinces had then, or have even now, been fully explored. By 2005, the view is firmer still: the world has been "very thoroughly explored".[96] What, then, one is bound to ask, have so many companies been doing for the past decade and intend to do for several more?

This model, the outcome it would seem of a restrictive geological world view, has failed. It is not difficult to itemise the reasons for this. There is the hard-line view taken on discounted Middle Eastern reserves, the exclusion of technology's role, neglect of the dynamics of government policy and realpolitik in the oil world, the static definition of YTF, little cognisance of recent trends in exploration effort, and the lack of sufficient attention to the business of geopolitical and corporate access. Above all, there is the lack of dynamics. There are, according to the model, no unknowns of significance: there is a *quantum fixe* and we know all that we need to know about it.

We can complete a more detailed picture of Campbell's work by considering, more briefly, some of his other texts. Perhaps the first to capture the public attention was a piece co-authored with Laherrère for *Scientific American* in 1998. In that widely read article the decline in the production of conventional oil was to begin "before 2010". There is the familiar dismissal of the potential of the frontier and future exploration, a key element of which is the view that the deepwater realm "has been shown to be absolutely non-prospective for geologic reasons".[97] This is not a view shared by all operators.

This view was at odds in 1998, and remains at odds, with the steady development of the deepwater game beyond the "Golden Triangle" of Brazil, Gulf of Guinea and Gulf of Mexico. Has anyone informed the Mauritanians or those companies now leasing the East African margins, Colombia, Venezuela and new tracts in the Northwest Shelf and elsewhere that they have been wasting their time? There is too a sceptical and conservative view of unconventional reserves (only 700 BBLS to be produced over 1998–2058). Likewise, Africa is said to have a "practical limit" of only 165 BBLS of ultimate recoverable reserves. Peak Oil here seems to have bought into the popular notion of Afro-pessimism.

By *Oil Depletion – Updated Through 2001*, Campbell's critique of the "flat-earth fraternity" of economists had become more pronounced. The corollary is a forecast of conventional oil production remaining flat until 2010 (although apparently the peak may already have passed in or soon after 2000). Deepwater oil is forecast to peak in 2010 at 8 MMBOPD. Its ultimate endowment is set at only 60 BBLS worldwide, with even this figure thought by the author to be optimistic.[98] Campbell's views by now on the social impact, relatively relaxed in the 1998 piece, have become more pessimistic: "Hydrocarbon Man" will be extinct by the end of the century.

Then *Forecasting Global Oil Supply 2000–2050* (2002) contains much familiar analysis, augmented by a critique of USGS methodology. The production plateau forecast earlier is now expected to be "anything but flat"– there will be price surges, recessions, tensions and growing conflicts. A peak for liquids (from all sources) is forecast for 2010 and a peak for oil-gas for 2015. "The historical pattern of economic growth has to end." By this time the scenario for society has become as important as the technical foundations in oil depletion – and it has deteriorated. In an interview of the same year, Campbell predicted that the effects of falling production will be "war, starvation, economic recession, the extinction of homo sapiens".[99]

Not dissimilar dramatic texts have continued to appear in a series of ASPO documents – notes, updates, articles, newsletters and presentations. The concern to "raise awareness" appears to have become dominant. Campbell as activist has been highly active, participating in 2004–05 in ASPO conferences; meetings with permaculture associations, community action groups, bio-energy associations, green parties in Sweden and Ireland, and Swiss pension fund managers; depletion workshops; and so on. Despite this busy agenda, the writings have continued. The *Final Energy Crisis*, published in 2005, included three anchor contributions, with a regional review that inexplicably fails to find place for Africa.[100]

The development of Campbell's career over recent years can be summarised as follows. He has retained his basic theory, thus providing (as we have seen) the paradoxical spectacle of a static, all-knowing model yielding ever-changing forecasts. To this has been added an increasing emphasis on the social and political aspects of our future, informed by an increasingly pessimistic and at times problematic vision. According to *Oil Crisis*, for example, not only is peak production imminent, but a bottoming out is just down the road (around 2020). Symbolic of the overall state of Campbell's thinking is the odd juxtaposition in that work of passages of technical exposition combined with others of messianic character. Overall, we may think that from the dry, rock-focused, ultra-conservative discipline of geology, a prophet has magically arisen.

This analysis of the major Peak Oil theorists would not be complete without an examination of the work of Campbell's sometime co-author, Jean Laherrère. With a rich career in the oil industry, he has worked for Total and CGG and consulted for Petroconsultants. As well as working with Campbell on Peak Oil forecasts in *Scientific American*, he has been involved with SPE/World Petroleum Congress (WPC) reserve-definition committees. He is a prolific author: Campbell's bibliography cited 32 papers and more have followed.

Peak Oil theory is in general fond of graphs, but Laherrère is a graphmaker par excellence. For him, "a graph is worth a thousand words".[101] Laherrère's graphical treatment descends even (and this is sometimes unusual for Peak Oil theory) to the level of representation of particular oilfields. Indeed, it is easy to feel that anything capable of two-dimensional representation finds embodiment in his images: graphs on North Atlantic cod landings, fertility and population, GDP and unemployment, debt and even happiness find a place in this extensive work. In general, however, the focus has been much more technical than Campbell's and the author appears much less concerned with social and political scenarios: his focus is consistently on oil.

Compared with others, Laherrère's treatment has been more mathematical and abstract. From an interest in the natural distribution theories of Pareto, Zipf and Mandelbrot has developed a fascination with the parabolic fractal. This is wedded to considerable statistical erudition, a heightened preoccupation with definition and a desire for numerical exactitude. Along with this unusual sophistication, however, comes a reliance on many of the canonical principles of Peak Oil doctrine. There is a disembodiment from corporate strategies and government policies: at best they are reduced to hidden dimensions concealed within statistical frequencies. The result is a theory of great technical accomplishment but real-world aridity.

Characteristically, the data and diagnoses of other parties – the USGS, Shell, the EIA and the IEA, to cite some – are often discounted. There are, apparently, three oil worlds: that of economists, flawed by their reliance on politically suspect data; that of managers-politicians, undermined by their obsession with growth; and that of technicians, who are blessed with access to real data but who tend to speak freely only when they have retired. Technical databases – and here we descend into the complicated *demi-monde pétrolier* – have been compiled in part, apparently, from data provided by "spy companies" (Petrologistics, presumably, among others).

Let us examine Laherrère's central thesis and some of its raw numbers. In 1996 as an associate of Petroconsultants he estimated the URR (the "fixed oil" stuff) in a range of 1,700–1,800–2,000 BBLS, noting that the average estimate for the previous 30 years had been 2,000 BBLS. Undiscovered YTF oil was estimated at some 150–200–400 BBLS (one of the nice things about numbers is that there are so many of them). He then forecast that the coming crisis "may arise within ten years", that is by around 2006.

By 2006, however, instead of a peak we have a new forecast, suggesting a peak sometime in 2012 or after (though with a mid-point already at 2005, the dissymmetry brought about by "multi-cycles"). Uniquely, however, Laherrère also provides a peak date for all liquids based on an estimate of 3 TBLS stock (and calcu-lated also for "an unlikely" stock figure of 4 TBLS, with 1–2 TBLS allowance for non-conventional oil). The lower URR-all liquids figure produces a peak in 2015–20, with the "most likely" outcome an extended bumpy plateau (2010–20). The higher stock figure would, of course, justify a later date.

There are some unusual departures from Peak Oil orthodoxy. The timescales have begun to lengthen, the peaks are less imminent and ranges have been introduced. Moreover, some admission is made of the relevance of demand adjustment, perhaps even demand destruction. It all begins to sound (at least to one observer) just a little reminiscent of that bête noire of Peak Oil, Odell (the supposed "flat-earth" economist), whose most recent work has cast a net across the 21st century to depict a unique vista for the future shape of world oil and gas.

We should not, however, push this deviance too far. Many Peak Oil shibboleths remain in place. There is, for example, the typically gloomy view in 2005 of the potential in the world's

THE BATTLE FOR BARRELS

deepwater oil, which it is argued has so far exhibited two discovery peaks but will have only one for production. It all rests on the Golden Triangle again, along with Angola. No consideration is given to the possibility that the deepwater game may be still quite young and widening across the world: four or five dozen companies involved in this world would be surprised to hear that the game is almost over.

Another shibboleth is "mistaken R/P usage". Economists are accused of saying, on the basis of using an R/P ratio of 40 years, that there is no oil problem. Or they are accused of abusing the measure for forecasting purposes. (BP has been accused of similar infringements of competence.) It may be agreed that Laherrère makes some interesting and correct observations on the R/P conundrum. This is, however, no reason to infer that a static R/P is some entrenched formula employed by economics. For that purpose, a dynamic R/P (with non-linear dynamics taken on both sides) would be required – a development that would appear entirely at odds with Peak Oil's static view of the world.

To an economist, Peak Oil's core model, especially its two-dimensional variants, are reminiscent of the once popular models of simple economic growth based on capital functions. Such models too were highly abstract, elegant and perfectly suited to two-dimensional graphic representation. They proved, of course, highly reductionist, lacking in explanatory power, and insecurely related to the complex empirical world. It is significant that economic theory and practice has moved on, leaving these models behind. Perhaps some of Laherrère's Peak Oil non-orthodoxy – the use of numeric reserve ranges, more distant production peaks, plateau-type outcomes – reflects a step in a similar direction. Even so, as the high ground in Peak Oil is well populated by a few notables there have been new entrants onto the turf.

12

Glitterati in the desert

Having examined the work of what may be called Peak Oil's version of the Holy Trinity – Hubbert/Deffeyes (a sort of composite character), Campbell and Laherrère – it is time to turn our attention to a more recent writer who, though far from conventional in Peak Oil terms, has been greeted with adulation by the movement: Matthew Simmons. His *Twilight in the Desert* took the oil world by storm in 2005. It offers a detailed analysis of the oil industry in Saudi Arabia: its reserves, problems in oil capacity and production, and ageing super-giant and giant fields (especially the world's largest field: Ghawar).[103] It engages trenchantly with all manner of complex technical considerations, especially the Ghawar water cut issue and Saudi Arabia's secrecy over its reserves. It comes to the conclusion that all is very far from well in the House of Saud.

Simmons's approach is far from superficial. It exhibits the mindset of an experienced investment banker. His research, supported by much citation and reference, involved recourse, for example, to almost all SPE technical work in over 200 reports on Saudi fields, the 1974 US Senate Subcommittee record of hearings on Saudi Arabia and the 1979 US Senate Staff Report. It also included a visit to Saudi Arabia as a guest of Saudi Aramco in 2003. There were subsequent heated debates at the Center for Strategic and International Studies (CSIS) in Washington on Saudi reserves and production with Aramco oil officials in 2004. The book's conclusions are important: a peak in Saudi oil would be a very serious matter indeed.

The essence of the case is as follows. While the US and the IEA (and Aramco too) expect there to be adequate productive capacity for the future (10–12 MMBOPD or even 15–20 MMBOPD), Simmons estimates now-demonstrated capacity at only 10 MMBOPD maximum. The Saudi government claims an ultimate potential of 461 BBLS in reserves and Aramco estimates that it could install a productive capacity cushion of 2 MMBOPD in the near term. There is now an investment programme to meet this goal. Simmons, however, is sceptical. He dismisses Saudi claims that the inventory of discovered but not yet productive fields, together with the potential for new discoveries, will be sufficient to sustain production of 12–15 MMBOPD for 50 years. Amid much secrecy on the part of the Saudis, Simmons believes that there is even a question mark over the reported 261 BBLS of proven reserves.

During the severe 1993–94 price collapse, a former Aramco employee of 30 years told the author that "the Saudis will need to learn to grow dates and ride camels". They haven't had to yet, but will they have to in the future? Simmons points out that 90% of Saudi oil is drawn from seven giant fields that have matured and are growing older by the day. Water injection is, he says, a major problem and the sign of imminent depletion. There is, Simmons argues, limited potential from YTF oil because the country has been explored intensively already. Exploration efforts in the past have hardly proved successful; unexplored zones are said to be small and the Saudis' estimate of 200 BBLS potential untenable.

The Saudis are, in Simmons's view, headed for the sunset. The absence over the past couple of decades of public field-by-field data and the lack of third-party audit arouses deep suspicions. The Saudis, however, defend their figures staunchly. Their reserves, they argue, are spread over 85 fields and 320 reservoirs. Their assumptions on ultimate recovery rates have been cautious. Their figures are if anything, they claim, conservative. They are based on an expanding knowledge base, sophisticated management, and well-established, strict methodologies. There is further potential from technology to apply. For Simmons, however, the bottom line is that the estimates for the real proven oil reserve numbers are dodgy. The Saudi capacity limit has probably been reached.

Simmons's scenario makes for great drama and it is unsurprising, therefore, that it has been acclaimed by the Peak Oil community. Simmons himself, however, appears far from fully committed to either Peak Oil theory or its predilection for doom and gloom. He even makes extensive use of economic analysis and recognises that a rise in crude oil prices (perhaps to over $200 per barrel) would make an impact. He envisages a world still "working beyond Peak Oil". Perhaps we will manage to execute a Plan B or even a Plan C based on a new energy miracle. After twilight in the Saudi desert we might even witness "the dawn of

GLITTERATI IN THE DESERT

a more enlightened and sustainable global society". It could all lead to "the dawn of prosperous and democratic new societies throughout the entire Middle East, Nigeria, Libya, Algeria and Venezuela".

For all the homework that he has done, however, Simmons's views are far from generally accepted. Jim Jarrell, president of Ross Smith Energy Group in Calgary and head of a well-known independent firm of petroleum engineers, offers a withering critique in *Geopolitics of Energy*.[104] Simmons's work is, argues Jarrell, characterised by technical "gaffes", a lack of context and a neglect of market dynamics. The technical concerns raised by Simmons are said to be unexceptional in maturing fields and certainly not evidence of any crisis. The supposedly "fuzzy logic" deployed by the Saudis in managing their fields is a probabilistic method of acceptability. The reserve assignment method they use is as conservative as those of public quoted companies. The methodology follows a clear constitution of sound principles. Indeed, 50% of the 260 BBLS are "proved developed". Enhanced recovery upside is excluded in Saudi classified proved reserves, so when fields are taken offline for market reasons the reserves are then reclassified as undeveloped (some 41 BBLS is reportedly involved). Based on a technical understanding of the engineering involved, Jarrell argues that the water cut issue is being well managed (with several arcane engineering reasons provided). At 35% (now reported by the AAPG to have fallen to 30%), the water cut is considered acceptable.

Jarrell suggests that Simmons has overestimated the intensity of exploration to date (69 exploration wells in the past ten years is not an excessive effort). With large reserves and excess capacity for most of this time, it would not in any case have made sense economically to go hell-for-leather for new exploration, especially since 23 developed reservoirs out of a sizeable inventory of 80 discoveries provide immediate installable capacity for the future. Jarrell even views the USGS-2000 "mean expectation" of 89 BBLS for undiscovered oil (estimated when oil prices were very low) as conservative. Moreover, some of Simmons's geological observations of significance are found to be mistaken. The Saudis have had, Jarrell tells us, excess capacity in their reserves for 50 years. There is no evidence whatsoever for an imminent crisis. There will be no early sunset.

Jarrell, it should be noted, is far from being a lone voice here. The detailed review of Simmons's argument provided by Lynch in *Crop Circles in the Desert* (2006) is overwhelmingly critical. His critique is not limited to the details of technical or economic analysis: it concerns the big picture.[105] Simmons's thesis, says Lynch, depends upon a "what if?" scenario that, although entertaining, is not realistic. Projections of the amount of Saudi oil "needed" have been consistently overestimated since 1996. The potential of small fields has been overlooked. The best estimates on Saudi proved reserves (assuming IHS here) come in at 294 BBLS – higher even than Saudi Arabia's own numbers and 100 BBLS above ASPO's. It seems we need no Plan B. So Plan C can also wait for now.

Twilight can mean many things: at the equator, for example, night can fall like a sheet of lightning, whereas twilight in the summer climes of the northern hemisphere is of longer duration

– longer still in the white nights of the Arctic. Yet the danger is that this particular, somewhat flawed, image of twilight will now be recast in stone as Simmons take his place in Peak Oil's rigid pantheon of gods.

I3

Onslaught on heresy

Little attracts as much criticism from Peak Oil thinkers as the notion of reserve growth and the related concept of technology enhancements in oilfields. Particularly suspect in this regard is the USGS, especially since the departure of Charles Masters as its head. The main focus for criticism has been its *Report of Year 2000*.[106] This is not entirely surprising, since its data are not consistent with Peak Oil forecasts. The report is said to enjoy "undeserved credibility" and even, perhaps, to be following a "hidden agenda". Its critics included Laherrère and Campbell and most of Peak Oil as well.

In his critique, Laherrère focuses on, *inter alia*, the USGS's estimates of undiscovered oil (939 BBLS potential over one-quarter of a century) and also of reserve additions (730 BBLS for potential field growth). Though he claims that these estimates are "excessive", it is not proven that they are fundamentally wrong. Moreover, he treats the undiscovered oil estimates as actual performance projections for new discoveries over a defined time – but these measures were never intended to serve in this way.

This critique of the USGS also underrates the potential for

further technology development. For example (to give just one), 3-D (or 4-D) seismic technology has yet to be applied ubiquitously (or even, one suspects, to 50% of world acreage). Laherrère also criticises the USGS's use of the Delphi technique (for its subjectivity) and its arithmetic regarding means, modes and probability distributions in Monte Carlo simulations. Several arithmetical criticisms are sound and others are somewhat pedantic, but overall they do nothing to undermine the broad thrust of the USGS's projections. Overall, the main source of Laherrère's unhappiness appears to be the USGS's revision upwards of global URR from 2300 BBLS in 1994 to 3,345 in 2000.

In a related instance, Laherrère also sent a letter to Phil Watts, then chairman of Shell, criticising his company's use of USGS statistics to exhibit incredulity that Shell seemed to have used these as a low estimate in its own estimate of ultimate reserve potential to 2100 put at 3.8–7.0 TBLS (the USGS, it will be remembered, was looking ahead some 25 years). We should note that it is not only Shell that sees an upside and sizeable stock of world oil reserves: in 2006 Statoil unofficially estimated world reserves at 1,402 BBLS, with exploration potential of 305 BBLS and reserve growth potential of 535 BBLS. Statoil's estimates of resources/reserves thus total 2,242 BBLS (with URR at 3,264 BBLS).[107] This suggests that further exploration alone could suffice for many years, or even that the world might do with just reserve growth. In all likelihood it will be a mix of both. Statoil also noted the worldwide underdevelopment and low investment in new fields in the past, and regarded field redevelopment potential as large and much neglected.

Campbell's critique came in response to "Global Petroleum Reserves: A View to the Future", a piece written by USGS geologists and published in *Geotimes* (2002).[108] Among Campbell's targets in what he regards as a "thoroughly flawed study" is its employment of the notion of a resource pyramid illustrating potential transitions from lower-cost oil to more costly oil (conventional or otherwise). Yet this is not unusual or out of line with what companies do: they work their portfolio options and optimise according to strategy and cycles as well as funds.

Similarly, the method of applying US onshore reserve growth in older fields to the international situation is dismissed as "absurd" on the grounds that different conditions obviously apply. This may not always be the case. The fact that actual discovery over the previous seven years is less than the original estimate of the amount available for discovery is taken by Peak Oil as evidence that the USGS model is flawed. However, the USGS was not forecasting time-specific discovery trends in the first place – only the potential of reserves to be found.

One cause of great concern to most in Peak Oil is the role of reserve growth. Given its importance, let us dwell on this concept awhile. A consideration of the way in which reserve creep dynamics come into play in upstream decisions helps to illuminate the way the oil game is played within local fields, basins and countries. Though Peak Oil thinkers tend either to exclude reserve growth (on the grounds that it is all just part of discovery) or to discount it (on the assumption that technology is, at best, neutral), for oil companies it does not work that way. They are intent on achieving reserve growth.

Of significance worldwide has been investment in capacity as a determinant of reserve access and future oil prospects. Companies have several options with regard to new reserves. They can add to partly developed reservoirs; develop known but undeveloped reservoirs; search for new oil reservoirs in existing fields; go for new fields in "old plays" or new plays in old basins; and search for oil in new basins. The industry as a whole acts as a giant scanner, seeking to discover and exploit reserves, all of which requires investment. The process of asset selection within the internal capital market of the company is essential to the investment allocation strategy. Development investment in aggregate often "provides identified new reserves", moving resources from both the probable-to-proven and the possible-to-probable categories.

Peak Oil estimations characteristically treat investment as an exogenous variable (that is, "outside the model" and therefore outside the mind), thereby ignoring what may be most important of all: the ongoing, long-term commercialisation of oil reserves. This occurs as new provinces, and new areas within existing provinces, are opened up. It can take years to know the real volumes of a new pool, field, area or province: it takes time and cost to generate anything like full knowledge. In many cases, new ventures are aborted before inception on the grounds of ex-ante, post-geological assessment based on economic calculations, fiscal expectations, or other strategy assumptions. Many potential fields are then not "found", booked or developed; thus they often "disappear" from public estimates of reserves. This has direct implications for dynamic reserve growth estimation.

Here it is pertinent to note that Peak Oil forecasters do not enjoy an undiluted view of the state or corporate portfolios that contain these internal and hidden assessments which their models logically require.

Company reasons for shut-ins and the "concealment" of reserves may vary: fields may be too small for commerciality to be declared at the time of discovery; other prospects may be given higher priority in the exploration portfolio; constraints on a company's exploration capital budget may force some prospects off the priority list; there may be insufficient field or basin infrastructure to permit immediate exploitation; expected returns may be affected by issues such as fiscal policy or available technology; or there may be shifts in company strategies to other objectives (for example, to other countries, gas deals, or production from other fields). If the shut-in well is not now declared commercial, it does not get booked. But not being booked does not mean that the potential reserves may not be brought in later. It is a category issue that the SPE schema captures well and which Peak Oil forecasts merely discount.

There may be several sources of reserve creep during the discovery-to-exhaustion life cycle: during discovery periods the initial NFW may be upgraded in a short space of time; gains from the appraisal period and development phases may again add to field reserves (a process that may repeat); there may be step-out well additions and new plays at deeper or different targets; gains from lower costs as technology advances may permit greater reservoir extraction rates; reserve additions may occur in mature basins or fields as infrastructure synergies allow for new tie-ins, satellites and previously marginal opportunities to be pursued; improving crude margins (related to the price/cost relationship) may make reserves more feasible to exploit; and recorded reserve growth may come later from improved fiscal terms. Reserve growth is not, therefore, solely dependent on new discoveries. Most such decisions turn on changes that cannot be predicted at the time of discovery.

State policy may also affect estimates of reserves in commercial and non-commercial categories, especially in zones of resource nationalism. This results in slack that could enhance reserves in future. Government policies may cause shut-in volumes to remain in-ground for a number of reasons There may be slow, or inadequate, fiscal adjustment; a lack of terms suitable to marginal fields; too much focus on the concerns of large operators; a lack of policies that encourage the flow of new relinquishments; and large, over-dimensioned block sizes that keep incumbent operators in control of prospects that could otherwise be picked up by more aggressive and cost-efficient players. These conditions work against the immediate release of shut-in potential. They may of course change – but such changes cannot be accounted for in Peak Oil models.

Increasingly, the management of resources/reserves, present and future, has become related to technology issues, new technology development and applications, and, perhaps most critically, the management of technologies. More and more companies are building strategy on the basis of technology competencies. Technology is now widely recognised as having a great potential effect on reserve categorisation, discovery, development, growth and production potential. It is likely to exert even more of an effect in periods ahead. Technology will thus widen the range of exploitable resources and reserves, as well as provide greater accessibility to larger oil reserves. Technology operates here as a sort of analogue for knowledge as well. Nonsensically, however, Peak Oil models in effect require that the fruits of future knowledge be treated once and for ever as a "reporting phenomenon".

As a consequence, Peak Oil models backdate any appreciation in new reserve additions (enhanced discoveries and/or field growth) to initial discovery date. Odell, among others, does not consider this a robust technique. The why and where of reserve augmentation is, however, largely immaterial: it is their existence and availability at a given time that matters most. Odell points out that Peak Oil discovery profiles are invalid for forecasting oil supply. Recently discovered fields have less time to go through the normal, lengthy, reserve augmentation process than fields discovered earlier, so comparisons are inherently flawed. This includes the measure of reserve quanta in relation to quite different periods between which production costs and market prices might differ. Peak Oil's treatment of reserve growth makes the past look more attractive, the future less benign.

Peak Oil's backdating methods are, therefore, problematic and flawed. Our current knowledge of resources is finite but has high potential for change. More will be known about the global resource/reserve base in the next decade than was known in the last. As the world oilfield (so to speak) matures, the knowledge base grows. Full oil maturation has still to occur in many places. Eventually, the world's oil provinces may become fully known, but initially only in certain countries and then later in wider regions. The differences between these processes tend to go unrecognised in Peak Oil theory.

An empirical observation from Statoil sums up well some of these considerations in investment, field redevelopment and future reserve growth potential.[109] It defines "outtake" as the annual production of remaining developed reserves and shows that an improved recovery outtake or reserve growth from remaining developed oil reserves estimated on 9,000 fields could generate 540 BBLS (even using only today's technologies). No measure is made for future technologies. This suggests the world is unlikely to run short of realisable commercialised reserves any time soon.

I 4

Peak Oil meets economics

Given the problems concerning Peak Oil theory, it is not surprising that it has been regularly criticised over the years. Most of the published criticism has come from specialists in the dismal science: economists and other analysts and commentators associated with the industry. Chief among the economists have been the doyen of oil economists, Maurice Adelman, the much-respected Peter Odell and Michael Lynch. To these may be added CERA's Daniel Yergin, whose background appears less purely in economics (his work has involved oil historiography, academia, consulting and energy strategy), but he nonetheless writes from a similar perspective to the others. For the purposes of shorthand, we may think of this group as the "economists".

The economists have taken issue with Peak Oil method-ology, especially its geological determinism. They do not agree with the conclusion either that the oil era is now over or that we will suddenly become beholden solely to Middle Eastern states controlling the remaining large reserves. Some non-economists have also challenged Peak Oil theorists. The worry in particular has been the risk that their apparently scientific work will exert an undue influence on (not necessarily well-versed) policymakers and media figures.

Lynch's critique is to be found in a series of detailed reports.[110] He argues that the Peak Oil theory – not least because its estimates of URR are too low – is alarmist. He highlights the perpetual reincarnation of forecasts, the opacity or impenetrability of many Peak Oil studies, and the pattern of errors and mistaken assumptions.

The use of the Hubbert curve by Campbell, Laherrère and others is a key focus for Lynch, as it is after all the source of much of Peak Oil thinking. First, the curve was applied to the Lower-48 in the US. Then it was applied to the world as a whole. Somewhere along the line it seemed to acquire an explanatory power; it was as if future production should follow such a curve (even if in many cases it did not). The model rests on the assumption that unrestrained extraction of a finite resource rises along a bell-shaped curve that peaks when about half of the resource is gone. But unrestrained extraction is not the norm in many oil societies, which is unsurprising given, for example, the impact of government policies on tax, regulation, and so on. Shifts in state policies can have an impact on production profiles. This undermines any idea of predetermined production patterns based on a supposedly "immutable physics of the reservoirs".

There is critique too of Peak Oil's apparent use of a proprietary database not available to the public at large. This, according to Lynch, provides a "convenient shield" against criticism. It also raises questions of intertemporal consistency and robustness (so even access to or use of this database may be no guarantee of the

Peak Oil claims). This issue is compounded by the view taken by several at IHS (Peter Stark, for example), its subsidiary-affiliate CERA (notably Yergin) and the USGS. Their URR estimates based on original Petroconsultants data are all much greater than Peak Oil's (by a margin of up to 1.2 trillion).[111] There are also, as discussed earlier, awkward questions concerning reserve growth.

Lynch notes that Peak Oil "evidence" is usually presented in visuals, with graphs, creaming curves, bell curves, and so on, sometimes for specific countries or known fields. Such visuals rely on a projection of the past into the future but often provide no explanation of why this method is valid and importantly why, as we have seen, so many factors should be held as exogenous. According to Peak Oil assumptions, these factors play no part on the future of, for example, exploitation strategies, reserve endowment, or production. The result is a model of a future that has no future.

Virtually all Peak Oil's visuals make claims of correlations without providing the statistical results or tests on which they are based, according to Lynch. Sometimes only a few data points may be found or inferred for the suggestion of rich, even complex trajectories. Any future discoveries of giant fields are ruled out of court: they are supposed to be discovered first. And yet this has not always been the case. Particular problems revolve around Peak Oil applications of the "law of natural distribution", where fields are ordered by size to yield an asymptote, so indicating an approaching limit. However, when fields are organised by time of discovery, this asymptote is often unclear. Here the model appears to be self-serving.

Lynch also criticises the creaming curve "fits" offered by Laherrère, arguing that they provide no statistical analysis of actual correlations: the so-called "fits" are imprecisely adduced without adequate quantitative data. A similar criticism is made of Deffeyes's Gaussian curves. Moreover, there is no defined lag between peak discovery year and peak production year, or any evident way of predicting one. This diminishes predictive power. In the circumstances the only inference that can be drawn is the unenlightening one that production peaks tend to come after discovery peaks.

The fundamental critique of all the Hubbertian models used by Campbell, Laherrère and Deffeyes, put forward by Lynch, is that they mistake correlation for causality. It is assumed that changes are caused by nature alone: falling discovery rates, for example, are seen as the result of natural scarcity rather than of, for example, changes in crude oil prices, capital market influences on drilling, or the nationalisation policies of the 1970s. Here could be added the downside in world exploration effort in the 1990s as a result of global corporate shifts in strategy towards exploitation.

Lynch has focused particularly on the Middle East, an area with large reserves that are discounted in Peak Oil. He argues that it has not been played out in geological terms. Actual performance has been influenced strongly by government policy on access and allowable activity. Laherrère's figures for NFW/discovered volumes in the Middle East for 1980 and 2002 are also noted as mere aggregates. They fail to pick out such intraregional influences as those induced by, for instance, the Iraq–Iran war, the sanctions periods for Iraq and Iran, and the increase in drilling in lesser states such as Yemen, Syria and Oman. Peak Oil theory has treated all these wildcats as somehow equal and it draws fallacious conclusions as a result. This observation holds implications for the future, since recent wells and expected wells are likely to be drilled with better technologies and knowledge applied.

As with others, Lynch has documented the relationship between Campbell's pessimism and his forecasts. He points out that changes in quantities always produce changes in the dates forecast, but never in the model or viewpoint that links them. Campbell's URR estimates have risen significantly (from 1,575 BBLS in 1989 to 1,950 BBLS by 2002). Over 1997–2002, Lynch points out, country estimates were also revised (the amount of discovered oil exceeded URR in 30 of 57 countries for which data were given). Lynch examined in *Crying Wolf* the detailed reasons why Campbell's estimates have turned out to be mistaken. Many errors are the result of the impact of national oil companies. Many developing countries' production did not follow the shape of the bell curve. Even in mature, open regimes such as Canada and the US, divergent trajectories have materialised. Yet somehow in Campbell's serial projections (1989, 1991, 1997–98, and so on) the ultimate peak is always imminent.

Here Lynch is not alone. Many share his non-pessimistic conclusions. Wood Mackenzie, for example, states that non-OPEC oil production has not peaked. It expects such output will grow until 2013 at least.[112] And to this peak if it then occurs must be added OPEC production and its large future potential. This observation is significant, since it is in this weaker non-OPEC oil world that Peak Oil thinkers expect to see the greatest decline. While Wood Mackenzie envisages a more tightly balanced oil market from 2015–2020 onwards, inputs from unconventional oil and alternatives are likely to shape new oil market paradigms.

Odell has also been a trenchant critic of Peak Oil forecasts.[113] In a detailed analysis of 21st-century carbon fuel futures, he estimates that in the case of oil the total resource base is likely to be as high as 5 TBLS (for all oil). Though conventional oil may peak around 2030, its effect would be offset by non-conventional output. The combined sources would peak only in 2060. According to Odell, the unconventional oil game has barely started. Volumes around 3 TBLS are feasibly recoverable from this resource. Non-conventional oil, he estimates, will not peak until 2080–90. Peak Oil forecasts are, therefore, undone – in strategic, market terms – by a consideration of interfuel relationships.

However, Odell expects the combined oil peak to arise at a lower than necessary level. This results in part from competition from other fuels, not simply from depletion. Moreover, the overall depletion profile is expected to be gradual, with the oil industry at the end of the 21st century being larger than when it started. Though Odell thinks that conventional gas may peak in 2050, a non-conventional gas peak is not forecast until 2100. Here Odell agrees with Adelman's view of the ultimate oil resource ("the unknown, the unknowable and the unimportant") but also adds long-term competition between fuels (especially natural gas) into the equation.[114] This is a reflection of real-world energy markets. He even considers the possibility, much discussed in Soviet and Russian technical literature, of inorganic oil. Without need for accepting such arguments, oil as a quasi-renewable fuel would, of course, open up radically new considerations.

Let us conclude this survey of economists' critiques of Peak Oil theory with one written several decades before that theory was even born. Writing as long ago as 1933, Erich Zimmermann (1888–1961) advanced the functional theory of resources to make many salient points. In place of resource fixity, Zimmermann suggested that man acts on resources to make appraisal as a means to extract supply.[115] Resources are dynamic, they become, and evolve out of the triune of nature, man and culture within which nature sets the outer boundary limits, while knowledge is the mother of all realised resources. Presciently, Zimmermann drew attention to the vast gulf in views between physical scientists and economists on resource questions. The former seek comfort in a proximate concrete reality and work with mindsets that emphasise the absolute, static and one-dimensional – all, according to Zimmermann, at variance with the real world. Since then the shift in focus in neoclassical economics towards econometrics and a pseudo-bounded universe, combined with the professional desire for "hard" mathematical relationships, has tended to obscure this set of astute observations found in economics many decades ago. Across the gap of more than three-quarters of a century, Zimmermann's observations hit home hard.

The economists' critiques surveyed confirm the overall view of Peak Oil theory that we have been developing: it simply is not possible to generate reliable forecasts of oil production from a static model built on an assumption of a *quantum fixe* and exclusion of such factors as crude prices, technologies, corporate strategies and government policy. The pessimistic forecasts that emerge from such a model are a result of the mindset and methodology of its designers.

Yet it is economics that has become known as the "dismal science". In view of Peak Oil's pessimistic geological vision, one might ask whether geology is similarly "dismal". Is there something about the mindset of geology that inclines people towards pessimism?

Over the past quarter of a century or more, the author has encountered a large, varied and often admirable cast of geologists, geophysicists, petroleum engineers and geoscientists of all types. Many were and remain employed in the world's best oil companies. They have included many from Petroconsultants: the in-house team of editors for the Foreign Scouting Service; managers, including some consultant geologists attached to the company, the former president, Gerry Dixon, and the former chairman, Harry Wassall.[116] Throughout that time, which included four years in the mid-1980s in charge of Petroconsultants' global economics and consulting groups, the author encountered little presumption of an "end of oil". In the 1990s and since, the geoscience fraternity at large encountered across the world has not held this view as domain.

The author's work for Global Pacific & Partners involves regular meetings with oil companies and state players worldwide. Over the years this has involved meeting at least 15,000 corporate executives and oil players, including many from the technical side of the business. They included geologists of all views and persuasions, vintages and skills, and with varied expertise. Few have agreed with Peak Oil's undiluted theory, and even fewer with its empirical claim to imminent peak. The same may be said of those corporate and government speakers from many and varied organisations who have given wide-ranging presentations at Global Pacific & Partners' conferences held in Africa, Asia, Latin America, Europe, the US and elsewhere since 1991.[117] Of these – and indeed of all those executives whom the author has met at industry gatherings in the far-flung oil capitals of the world – only a tiny minority would feel at ease with the Peak Oil mindset. Similarly, managers of new ventures – the cutting edge of the oil game – tend to hold optimistic views on reserve potential and hence oil supply, as do most independent players in the oil world who plough their dollars into the dirt.

Those from the top echelons of the geological profession, armed with state-of-art skills and technologies, focus on the potential of new and formerly closed areas including, most notably: Africa (especially Libya, Sudan, Nigeria and Angola), Colombia and Brazil, open zones in the Middle East such as Yemen, South-east Asia and Russia, as well as frontiers elsewhere. They pay close attention to several facets of the non-technical game: prospect potential and measured upside in commercial ventures, reserve growth expectations, the risks and uncertainties entailed in deal contracts, the economics of future investments, fiscal conditions, government policies, and so on. So there is no dismissal of dismal science inside the world's oil companies and among its existing geoscience practitioners.

As modern geologists, these people are required to be masters of many games. Most will have technical defaults to their name (the dry holes come from somewhere); several are the grand-fathers of some of the best discoveries. In general, what they have in common are convictions at odds with Peak Oil theory, namely that the above-ground obviously matters and that there is no imminent peak in oil production expected.

Part 4

WORLD OIL: COMPLEXITY AND CIRCUMSTANCE

It is time now to consider diverse viewpoints from within the world oil industry – those of what we might call the players. These include super-majors, large independents, minnows and start-ups, national oil companies and governments. All have a stake in the world oil future. All would be affected if imminent Peak Oil were to occur. Few of them appear to expect it to do so.

A convenient starting point is the question of the price of crude. Peak Oil theorists tend to see the movement of prices above the once-unforeseen level of $75 per barrel, reached during mid-2006, as evidence of the peaking and imminent decline in oil supply. Prices, they expect, will go even higher. They have in fact fallen sharply. Their view contrasts sharply with those of the players. Leonardo Maugeri (senior vice-president, corporate strategies and planning, at Eni Group), for example, explains the rise in prices as, more mundanely, the result of a mix of complex economic processes.[118]

The upward trend in the price of crude is not guaranteed to

continue indefinitely; and a significant 30% downward correction in late 2006 shows its cyclical potential. Crude price levels are a result of a combination of contemporary issues: resurgent resource nationalism that has "locked out" corporate oil; underinvestment in exploration over the past decade or so on the part of large and risk-averse companies; OPEC delayed responses (following overproduction in the 1980s and the 1998–99 price collapses) to the need for new capacity investment; sanctions against petroliferous countries in the recent past (Libya, Sudan, Iraq, Iran); a spare capacity shortage of approximately 2–3% of world demand (down from 15% in the mid-1980s); and a worldwide refinery capacity squeeze.

The rise of national oil companies on the world stage has not necessarily helped the pace of world exploration: most state players have augmented asset purchases, not initiated unbridled exploration. Like the oil tankers themselves, it takes time for the oil industry to respond to changes in such trends. Cyclical patterns and volatility inevitably result, very different in profile from the smooth depletion curves of Peak Oil theory.

If Peak Oil theory is from Mars, it seems that the players (like the economists) must be from Venus. This makes the apparent lack of a direct interface between ASPO and the industry an Achilles heel for the movement. The lack of engagement with real-world dynamics (notably today's exploration industry) helps sustain a backward-looking view of the world on the part of Peak Oil theory. Engagement might, of course, prove uncomfortable by making problematic the model's convenient exclusion of real oil world above-the-ground realities.[119]

15

Crude prices and crude reserves

The price of crude is a recurrent theme in any discussion of oil. We need to focus on this more closely and examine some of the influences on price changes and their implications, especially for reserves and company strategy.

First we may note the impact of the energy security premium. The oil price has recently been higher than it would be if it were determined purely by fundamentals. This is because of energy security policies based on threats, whether real or imagined, to crude supply. There are risks in access, capacity, production, trading flows and market efficiency. Many factors contribute to the resulting premium, including considerations of ideology, strategy and economics. Acquisition by state players as governments and national oil companies seek to own and control more oil abroad as well as at home has become a factor. It is difficult to be certain of the size of the premium. When prices in 2006 were in the range of $75–80 per barrel, many players and traders estimated the security premium to be at least $10–15/barrel. Some thought it might be higher. There is no way to be certain.

Another pressure on world oil and gas supply (and hence an

upward pressure on prices) arises from the value chain required to deliver oil to the world's consumer markets. Years of underinvestment have taken a toll. Neglect of energy infrastructure (for example, shortages of refineries in major markets, insufficient pipelines, non-synchronised installation of LNG terminals and limited capacity including shipping) has created bottlenecks and squeezed energy and product supply.

As we have seen, Peak Oil theory tends to assume that prices above $20/barrel have little or no effect on oil reserves or supply. This is a drastically simple-minded assumption, as can be verified by considering some of the effects of price rises on the world upstream industry. It is also a view, paradoxically, that exhibits a revealed price forecast ($20/barrel) independent of actual price – and hence one that for some considerable time has been wrong.

A trend towards higher prices augments the capital available for exploration. Recently, exploration companies have indeed floated on the AIM in London and in other capital markets such as Australia, Dubai, Calgary and Toronto. There has also been activity involving private equity houses and conglomerates such as Mittal (with ONGC) entering the upstream industry. Ultimately, the effects of an upward trend in price feed through into global exploration strategy, discovery effort and in the development of new reserves. But this is a long process of adjustment and not instantaneous.

The effects of crude price rises on reserves are multifaceted. They include, *inter alia*, the switching of reserves from probable and/or possible categories to a more commercial proven status; the recategorisation of contingent reserves into higher categories; an increase in exploration budgets and prospect targets in companies' portfolios, with in due course the development of new reserves; the opening of plays previous defined as submarginal becoming commercially viable; new frontiers attracting greater activity (especially now in Africa); the search for deeper targets, both offshore and onshore, becoming more attractive; the expansion of deepwater acreage, especially under exclusive economic zone (EEZ) provisions for littoral states worldwide; the expansion of economically viable areas extended within established plays; and an increase in corporate value (and thus the financial capacity to explore new ventures) occasioned by an increase in asset valuations for existing proven reserves. All this, of course, becomes discounted if it is assumed that increases in the price of crude have no effect.

Like the weather, oil prices are unpredictable. It is notoriously difficult to forecast them in the short term, let alone even 10–20 years ahead. We might, however, note some views on the future. Commentators put forward a whole range of price expectations during the rising price era in 2005–06. In July 2006, when crude was $78/barrel, Jim Rogers (co-founder of Quantum Hedge Fund with George Soros) said that crude prices would rise to reach $100/barrel. Others had talked of a super-spike in crude prices. Instead they suddenly fell within several weeks to below $60/barrel by October 2006, partly because of a reduced security premium as the US and Iran went into diplomatic negotiations.

Some industry players (the chief executives of Shell and BP among them) expect a lower future crude price regime to emerge. It may have already begun. These insiders draw on market fundamentals. This lower oil price world is expected to result from, among other factors, recent investments and planned field developments increasing the supply of crude in several key countries, notably in the Caspian region and in West Africa.

On the basis of historic long-term trends, Odell plumps for real crude prices (with a 2000 baseline) in the range of $18–22.35/barrel during the first two decades of this century. This case may have been superseded by events. It cannot now be said what will certainly occur. Whatever the future, it is extremely hard to predict oil prices. But it can be predicted that very high sustained prices would affect reserves and exploration.

According to Peak Oil theory, however, oil production will decline drastically over the period to 2020 – yet the price rises that might reasonably be expected as a result are assumed to have no upstream impact. Instead of, for example, an increase in exploration efforts or in capital invested in the global oil search, the effect of higher prices seem in Peak Oil theory to be negated by some sort of collective corporate paralysis. This is most unlikely. The sort of worldwide oil production downside envisaged over 2010–20 by Peak Oil's consensus would have a dramatic impact on the crude oil market, exploration and prices. This would induce more exploration ventures as well as radically affecting company strategies. It is also the case that past price crises – both on the upside and downside – have influenced industry efforts and strategy. So although nobody knows what the price of crude will be at a given future date, whatever the price, the *quantum flux* of oil reserves and supply will be sensitive to it.

16

Yesterday's oil world and past discovery

Peak Oil theory has paid attention to declines in global discovery rates and, indeed, uses them as a basis for forward projection. There is, in effect, an assumption that the world is virtually "played out". This view is largely ahistorical: it fails to attend to the particular events and contingencies that make up the recent history of oil discovery. Before using the past as a basis for forward projection, any theory needs to consider such factors as investment patterns alongside corporate and state strategies. It also needs to place them in a context of a shift in scenarios from Cold War "stability" to what may be termed "fractured globalisation" today, in which lack of access often denies companies their best portfolio choices.

Sanctioned states (including, at times, Libya, Sudan, Iran, Iraq, Myanmar and Syria) have been largely "off the table" for long periods for many players, especially those from the US. Even in today's Russia access has not been undiluted (especially for exploration, rather than selective acquisition). There has been less untrammelled entry worldwide to key oil states over recent years.

Harsher conditions in Venezuela have made it tougher to justify more new venture entry or enhanced investment. Even in some more mature zones and frontiers there have been political crises – in Indonesia post-Soeharto, in Aceh and other provinces, in Bolivia, Somalia, Liberia, Sierra Leone, the Democratic Republic of Congo (DRC) and elsewhere. The world is now replete with failed states (Somalia being the unenviable classic) and more failing ones. Many companies have regarded the political and commercial risks in Nigeria as high even if more companies have now entered this zone. Resource nationalism in the 21st century has added to difficulties across parts of the developing world – the Chad revocation of the Chevron/Petronas rights in the Doba oilfield is an example, even though they have now been restored by renegotiation.

In the broad sweep of recent upstream industry, Asia was the hot zone of the 1980s, followed by Latin America as it rapidly opened up (notably in Brazil and Argentina) in the 1990s. Africa too has become a hot zone since the mid-1990s. The Asian crisis had a negative effect on the South-east Asian heartlands (notably Indonesia, still paying the price for political instability). Now several Latin American countries have followed Venezuela along the road to state control, the onerous "state takes" that have resulted making it difficult for companies to justify investment in those areas. They have instead focused on new opportunities in Africa including, notably, Libya (once sanctions had been lifted). Some parts of the Middle East (notably Yemen) have also benefited, although core zones in that region (Saudi Arabia, Iran and Iraq, for example) are inhospitable to untrammelled exploration. The situation remains in flux: new frontiers are opening up. There are signs of an Asian resurgence (witness new bids made in India); Latin American countries such as Brazil, Colombia and Peru have attracted many independents; and Africa, despite some difficult countries, has become a particularly dynamic exploration and production zone.

The late 1990s saw a critical shift in the strategic priorities of the largest corporations (the "super-majors"). Against a backdrop of low crude prices and pressure on cash flow, exploitation gained in importance at the expense of exploration. These companies had been responsible, of course, for a large proportion of the capital invested in world exploration. The super-majors also became more interested in diversifying their portfolios and investing in, for example, development ventures, gas-LNG, GTL and unconventional plays within safer OECD countries. The larger independents, some of which were taken over or merged, adopted similar strategies to the super-majors. Since then, some of the smaller independents have been active in filling the gaps in the upstream industry and several new companies have been formed, especially targeted on Africa.

It is only recently that national oil companies have developed significant global, regional or overseas-oriented strategies. Most of the related investment has occurred since 2000. Those such as JNOC (now Jogmec), Pedco-KNOC and CPC-OPIC in East Asia that had formulated a world upstream strategy earlier were relatively small in risk capital commitment and did not meet great success in exploration. But this too is changing. The aggressive, fast-moving overseas strategies now exhibited by state players (for example, Petrobras, CNPC-PetroChina, CNOOC, Sinopec, ONGC, PTTEP, Petrovietnam, Pertamina, Sipetrol, Sonatrach, PetroSA) have focused on acquisition and reserve purchases. Some have also rebalanced their portfolios towards gas-LNG and unconventional options. Exploration abroad is becoming a more important priority for such companies, though sometimes at the expense of domestic exploration. For none of these players is exploration the top priority, especially for overseas ventures.

The result of this recent history is that there still remain oil zones with high prospectivity and existing fields that have not received sufficient investment. Let us see how this history is reflected in the numbers of NFW wells. Jack Kerfoot of Murphy Oil has shown how the number of NFW wells has been anything but constant.[120] There were 69,072 NFW wells drilled in the period 1980–84; this fell to 42,282 in 1985–89, then to 25,891 in 1990–94 and 21,326 in 1995–99. In 2000–04 NFW wells reached a low of 20,186. This shows the dramatic fall in the world exploration effort. It mirrors the new exploitation strategies of the supermajors, the shift in capital funds to development ventures and the emergence of state oil companies into the upstream world.

Since the early 1970s there has also been a shift in production from corporate oil to national oil companies. The top eight corporate players were drilling around 1,000–1,100 NFW wells a year during the 1970s compared with 150–175 a year now. Overall there was an 80% decrease in NFW drilling by the majors from 1975–79 to 2000–04. Nowadays, majors account for only 6% of NFW wells; independents account for 82% of the much reduced total. They do not, however, have large capital available to risk in exploration and replace the efforts made before. The exploration game and capital allocation within it have, therefore, changed dramatically. This will have had an impact on the recent discovery profile and will continue to do so (though none of this is to do with potential reserve non-availability *per se*).

As a glance at world basin maps or at drilling density records will confirm, many areas have yet to be fully examined. In Africa, the Cuvette and interior of the DRC is mostly untouched, as are interior basins in Angola, the large offshore and coastal zones in East Africa from Somalia to Mozambique and the Rift basins in Central and East Africa. The same is true of much of interior Brazil and many zones in the Andes and the southern reaches of Latin America. Even significant areas of offshore and deepwater Austral-asia (including New Zealand) have not been fully explored. The same is true of the vast landmass of Russia and its offshore zones in the Far East and the difficult Arctic zones. Hence there is much work for geophysical companies such as PGS, TGS-Nopec, Fugro and CGG in conducting seismic surveys to open new prospectivity vistas in newly defined basins, old zones and immature areas.

Global exploration is, then, clearly nowhere near its limits: much of the world remains "under-rigged" and chronically underexplored. There are a reported 130 large undrilled structures in Iraq alone. Deep horizons provide a further opportunity. Moreover, digital technologies are at an early stage in the world oilfields at large and could have a considerable impact over time on net recovered oil. At the moment, however, the level of drilling worldwide remains at historically low levels, and a significant upturn in overall exploratory drilling has yet to occur.

It is not exploration alone that has been affected by the recent history of the oil world. In the past few years (especially 2000–06), the world oil industry has experienced a number of co-terminal crises that have affected jointly access, exploration and production. They have served to diminish access to open acreage and reserve exploitation in some key oil-rich countries, thus reducing supply and adding pressure to the price of crude.

Venezuela, for example, has become more hostile to new upstream investment through various pre-emptive tax initiatives. PDVSA's strategies have been much compromised in terms of the oil production profile forward for 2006 (by up to 2 MMBOPD compared with projections PDVSA made in 1999). Iran has been under sanctions and is far from benign for many potential players, especially American ones. Iraqi oil production has been damaged by conflict and sabotage. Output for 2006 was, at best, perhaps 30% of real future potential. The conflicts in the Niger Delta in 2006 led to production losses of up to 800 MBOPD at one point. Disruption has continued (though at lower levels) as a result of a low-intensity civil war that has raged for years. It may continue for longer. There has been sabotage and hostage-taking, as well as threats to companies such as Shell, Chevron and Eni. In Indonesia debates continued for several years on terms and conditions for the sanctioning of the giant Cepu field, which will now, after contract settlement with ExxonMobil, produce at 170 MBOPD. There have, of course, also been the after-effects of natural causes (hurricanes among them). And in August 2006 Prudhoe Bay was shut down because of pipeline contamination and 200 MBOPD were taken off the market for a short time.

As a result of these difficulties in the recent history of oil there has been supply stress. Capacity in some producing areas has been stretched and the price of crude has been driven upwards, later to soften for specific reasons. The crucial point to grasp, however, is that none of this is geologically determined. The difficulties are serious but they are, as we have seen, readily explicable in terms of above-the-ground developments.

17

Governments and national oil companies

The present real or perceived paranoia based on expectations of ultimate oil scarcity comes at a time of mounting resource nationalism, concerns about oil security and complex operating conditions in countries such as Iraq, Venezuela, Russia, Iran, Saudi Arabia (and, indeed, the wider developing worlds of Africa, Asia, Latin America and the Middle East). Governments are in a position to modulate these conditions whether or not they chose to do so.

The illusion of an immediate gain to be made from realigning the historic state–corporate balance has encouraged resource nationalism. Bolivia is one example and it has, predictably, ended in grief, with a reversal of policy and, in the meantime, the loss of opportunity in terms of gas exports and even LNG. Resource nationalism makes access harder, though it does not diminish either resource or reserve potential.

In many countries the state has taken greater control of the oil game, diminished corporate entry options, imposed tougher terms on contracts, raised taxes, reversed reforms and in some instances renationalised or confiscated upstream assets. Corporate oil has had to adjust fast to these new global realities: it has invested in oil recovery technologies, widened exploration into new frontiers, and upgraded portfolios with a mix of gas-LNG and unconventional plays (such as heavy oils, oil sands and oil shale ventures), while also making more deepwater acreage commitments. In effect, corporate oil has been locked out of much of the high-quality world resource. The direct and indirect costs for the industry and the world are not inconsiderable.

Conditions for reserve access and replacement have become tougher, but not all developments upstream have been for the worse. Although governments acting as gatekeepers to oil reserves are often not fully efficient, many have rapidly improved their management of the upstream and exploration industry. There is much scope for further improvement in the future.

This geopolitical flux is passed over by the Peak Oil model, which simply aggregates countries together. In the process it ignores the potential for different developments in different places. The upstream world is highly divisible: states can act on individual terms, and there is competition in the primary acreage and secondary asset markets. Developments in the acreage market since 1990 are a case in point: more acreage is available now than ever before, but more of it is frontier acreage. Every year, more countries are coming to market and more blocks are offered for bid. Yet much of the known resource base with proven reserves is not accessible to most private players.

An example of global industry dynamics is the emergence of states with independent licensing for acreage and assets: in South

Africa (Petroleum Agency SA), Brazil (ANP), Peru (Perupetro), Colombia (ANH), Mozambique (INP), Algeria (Alnaft), Mali (Aurep), Indonesia (BP-Migas, to a lesser extent), Thailand (DMR), Jordan (with the restructured NRA), and so on. More upstream regime changes like this can be expected. The shift has encouraged exploration companies and new ventures. Most of this development has happened within the last ten years. It will take time before its effects are felt.

Many ministries have sharpened their strategy and game. There has been a flurry of roadshows and promotions (on Africa, Asia and Latin America in particular) in the key centres of the oil industry. None of the governments or agencies regards their potential as minimal, whatever Peak Oil theory may assume. Governments that have persisted, such as Uganda, have reaped rewards in terms of new oil discovery. Many more can do so as a new exploration paradigm stimulates a revision of opportunities and deal options across the world in Africa, Latin America (especially Brazil, Peru, Colombia), Asia (Papua New Guinea and elsewhere) and the Middle East.

Nonetheless, many governments still lack technical and other needed capacities and are able to open up only a few areas at any one time. A sizeable part of the world's acreage, especially in developing regions, is either non-licensed or has yet to come to market. Many national oil companies have typically experienced structural, managerial, financial, technical and operational weaknesses. In particular, there has been under-exploration as a result of privileged state oil company positions, quasi-monopolies, and first-taker preferences in the form of state holdings of the best acreage or excess assets. The result, in some cases, has been underperformance compared with that of corporate oil. In some instances, this has translated into less than best practice, when compared with corporate oil, in the spheres of exploration and field management.

Beyond this, there exists much inefficiency in terms of management and systems in many ministry-cum-national oil company entities and interfaces. This slack in the world system represents an inefficiency that may in future affect resource opening and reserve accessibility. Many countries are aware of such deficiencies and do introduce new contracts, tax regimes and fiscal adjustments. Not all have been benign for corporate players in exploration or development contracts. Algeria has recently reversed a draft law that would have created a more level playing field with Sonatrach, adding on windfall taxes at the same time. In doing so, it backtracked on an earlier reform agenda.

Though such initiatives seek more state revenue in the short term, they may affect critical deal flow and even thwart investment commitments in the long run. These effects can be significant. In particular, they heighten the competitive options. For example, in Papua New Guinea, Colombia and Brazil positive change has led to a surge in exploration and new oil developments after a lag. For the majority of African countries – even large reserve holders such as Nigeria, Algeria and Libya – the long term will matter most, which is why the majority will remain open for business. This is especially so for pure exploration destinations seeking their place in the hydrocarbon sunshine.

There is a problem with uneven playing fields. Take, for

example, the way that Nigeria's decision in 2005–06 to mix upstream bids and awards with mandatory downstream commitments (refinery, power and infrastructure investments) skewed the playing field. Many companies could not bid on such terms, and others did not wish to do so. Many bids received had later to be rescinded for non-fulfilment. There was also much spurious bonus bidding, with many firms unable to meet bonus cash calls. And it may yet turn out that some of the non-upstream deals do not get executed. This experience may restrict near-term performance in future. In the longer run, Nigeria is likely to learn the lessons. Such bid innovations are inefficient and are unwelcome in the industry. The Nigerian scheme was not in itself an example of resource nationalism, but it had similar effects. It reduced acreage access and corporate exposure while introducing a bias into the market allocation process. That the Nigerian government has had to reissue several blocks proves the point. Nigeria may have also paid a price in terms of the corporate quality of its new entrants.

Another example of state intervention leading to inefficiency is PDVSA under the Chavez government in Venezuela. Oil production has fallen below that projected in the pre-Chavez profile.[121] There has been underinvestment, inefficient fund allocation (with funds diverted to social and political programmes), less efficient management (with 20,000 senior staff fired in 2003) and less-than-best strategy. The result has been a deceleration in the exploration potential and in what could have been exploitation of Venezuela's vast resource and reserve potential.

Detailed reports have documented how many national oil

companies fail to operate best practice.[122] This does not mean that they are always "last in class", but it does mean that, taken together, they introduce systemic slack into the industry. As the state players control more and more of the world's oil reserves, so this phenomenon will potentially count more. But it has nothing to do with fundamental reserve inadequacy or some restricted geological determinism.

Fashions also come and go in the upstream world. The recent surge of resource nationalism and aggravated statism may prove to be part of an as yet incomplete cycle. If crude prices continue to shift down from the highs of $75 per barrel reached in early 2006, states that go too far in asserting claims will in time become less competitive. Then a new boundary line is likely to be set between accessible and restrictive oil regimes.

In most countries the exploration industry would be best served by sustained competitive contract terms with bid rounds and acreage awards that reflect normal conditions and practice as measured against world-class standards. Acreage leasing successes are not guaranteed. Crude price cycles can, and normally do, reverse. There could in future be large-scale openings that would radically alter the focus of corporate portfolios. Consider, for example, the impact that a stable regime in Iraq, a change of regime and a reversal of policy in Venezuela, openings for independents in Russia, or even a de-sanctioned Iran could and would have on the allocation of upstream capital and ventures.

From the purist Peak Oil perspective, these observations on recent oil history, players' responses and the price of crude would be irrelevant. They fall foul of the central, static assumptions of

Peak Oil theory – that below the ground there is a *quantum fixe* of oil, known once and for all, and the dynamics of the above-the-ground world are of no account. As we have seen, the relationship between the oil industry and government is indeed a crucial and dynamic one. In a world where state players hold 90% of world reserves and where the growth of that control has had a direct (negative) impact on rates of exploration and reserve growth, as indeed have tax regimes (and this, remember, in a world much of which remains unexplored), any theory that excludes such factors from its calculus must be found wanting.

18

World oil reserves
and future potential

Peak Oil theory rests on certain key numbers, especially for reserves (so much else having been excluded by the model and its assumptions). It is important therefore to examine those numbers and assess their validity and reliability. Hence it is crucial to look particularly at the issue of the resource and reserve base.

The world industry is distinctly out of tune with Peak Oil's song about its numbers. Leonardo Maugeri of Eni, a world-class company, estimates that there are now some 2 trillion barrels of recoverable reserves not yet classified as proven that will in time become commercially exploitable as technology and subsoil knowledge improves and higher prices improve margins. This is a significant volume. Also, according to Maugeri, only around 35% of known oil in fields has been recovered (up from 22% in 1980). The consensus within the industry is that this recovery figure may well increase in future, and higher recovery will be achieved over a wider spectrum of fields. There is too a vast stock of unconventional oil on which to draw in places such as Venezuela, Canada, China, Russia, Brazil, India, Madagascar and even the US. The Peak Oil critique of Middle East reserve adjustment in the 1980s is also viewed by Maugeri as explicable and justified due to new technologies and changed conditions. The Persian Gulf is regarded as a virgin zone in terms of exploration, and Iraq is considered to have probably about 200 BBLS more in reserves.

Moreover, Maugeri points to a discovery trend reversal over 1999–2003 in which time significant volumes of oil (at 60 BBLS) were reported as discovered.[123] Although gigantic fields in future may be fewer, the discovery trends may make this less relevant as the pattern of supply will be drawn from a mosaic of many different supply streams that will flow from a wide mix of sources including smaller fields and non-conventional oil. There are huge areas of the planet that have yet to be thoroughly explored. As Maugeri observes, only 30% of known basins in the world now produce oil and gas and many need exploration and modern technologies, and another 30% of global sedimentary basins are still unexplored. At present, subsurface knowledge of these basins is imperfect, scattered and often limited. It is thus apparent that the view of industry players is also out of step with Peak Oil's theory, in regard not just to its numbers but also to its concept on how to view the future.

The conceptual and empirical divergences between Peak Oil theorists and industry insiders are pronounced. The former's presumption of a fully explored world cannot be squared with the myriad upstream efforts and strategies of a slew of independent companies around the globe. Peak Oil adherents seem to have little appreciation of or exposure to the wide mix of modern independent players. It is unsurprising, then, that within the industry Peak Oil theory is widely regarded as alarmist or catastrophist, prone to a pseudo-scientific fatalism camouflaged in sophistic-ated models, with a record of erroneous forecasting that is likely to continue.

Much of the difference in view hinges on the question of resource endowment and reserve estimation. Odell has surveyed estimates for the oil resource endowment that stretch back over 70 years or so. In the 1940s the world was thought to have ultimate reserves of less than 730 MMBLS. By the 1970s the estimates had settled around 2,200 MMBLS. Since then the search for oil has not just continued but intensified and extended geographic-ally. Industry research on exploration and production continues to show systematic results. Shell and others have subsequently estimated 500–1,000 MMBLS as YTF beyond the existing reserves in place, with another 400–500 extra potential in reserve growth (this is close to the USGS-2000 view, with its estimate of 3,000 MMBLS in total for the conventional oil endowment).

Apart from these numeric variances, there is a clash of analytics. Against the finite, static Peak Oil model sits the dynamic perspective of the players. The latter signal that both reserves and production are inherently dynamic over time. A non-linear view would hold, moreover, that the true stock/flow balance is found to be dynamic in the ultimate resource/reserve ratio (shown in the SPE classifications). On this view, there is potential for significant future impacts relating to reserves and production, tempered by the evolving and unknown constraints on oil access. Players see oil production as elastic (though not perfectly so). For Peak Oil theorists oil production is a zero negative game; for players it is a zero positive one.

The fact that Peak Oil takes the minimalist road to reserve estimation builds an inherent bias into its model that is of concern, especially given that the industry as a whole takes a much more realist and optimistic view. It is therefore instructive to look at some of these differences and the reasons for them. Though it is obviously not possible here to survey every reserve position in the world, it is possible to provide an overview. We shall begin with Africa, a vast and in many places frontier world.

The Global Pacific & Partners 13th Annual Africa Oil Week and Africa Upstream Conference in Cape Town in 2006 (15–17 November) attracted, as always, the crème de la crème in government, national oil companies and corporate players from around the world, all of them keen to assess the African upstream future in oil and gas-LNG. The annual conference provides the world's best insight into the upstream and exploration industry in Africa – past, present and future. Since 1994, when Africa Upstream was initiated, consistent African upstream reserve growth and oil development have been recorded. This has been driven by growth in the scope and depth of exploration, state strategies, multibasin openings, a radical upsurge in bid rounds, mega-investments and commitments, and a multiplicity of new ventures by corporate entrants, especially new independents, from six continents.

All this contrasts dramatically with the apocalyptic images of Africa portrayed by Peak Oil's thinkers. They present Africa as having limited oil resource endowment, modest reserves and a constrained oil production outlook – not at all what has been observed by the 5,000 or so senior executives and government officials who have participated in Africa Upstream over the years. The idea that nothing would change in Africa is obsolete. Many companies have targeted Africa for new ventures and development, and oil production has grown fast. The facts on and below the ground in Africa belie the pessimistic Peak Oil theory and numeric "truths".

The Peak Oil claim in 1996 was that Africa had about 11 BBLS of oil in the undiscovered category. In 2001 the claim was that Nigeria had just 1.9 BBLS YTF, Angola a paltry 1.1 BBLS, Algeria and Libya 3 BBLS each and Sudan 0.5 BBLS. This left 1.5 BBLS for the rest of Africa, comprising a further 49 states. Many of these states were, we should note, already oil producers, most with ongoing exploration, many with prospective frontiers. There is a vast set of marginally explored inland basin zones, as well as enormous littoral offshore areas and deepwater (including EEZ) acreage at an immature stage. Oil reserves have been found and discovered much in excess of Peak Oil's limited estimates. Discoveries made have been well ahead of the YTF endowment presumed by Peak Oil estimates, which vary hugely from the best-available African oil reserve prognostications.

Global Pacific & Partners makes country-by-country estimates for Sub-Saharan African oil, with the usual caveats, based on diverse industry sources, including best selected official data, company measures and independent geoscience estimates. The estimates focus on proven oil and potential, implying best estimates for new oil reserves that might be available for discovery (whether by 2025 or beyond). Guidance was sought from operators and governments, as well as national oil companies – they should know something at least. So let us consider the African oil producers, the exploration-prospective states and the frontier players.

There are now more and more oil producer states in Africa, and more will join this list in time. In Angola, there are proven reserves of around 15 BBLS. Sonangol has estimated ultimate potential at some 50–70 BBLS. Cameroon's proven oil may total only 200 MMBLS, though others cite 400 MMBLS, with future potential estimated at 300 MMBLS (excluding deepwater potential to be considered). In Chad, 1 BBLS is already proven and there may be 1–2 BBLS awaiting near-term confirmation and discovery. Congo has proven oil at around 1.5 BBLS, with potential for another 1 BBLS expected – much depends on the deepwater plays. Discoveries in Côte d'Ivoire total 500–1,000 MMBLS, with the government claiming (perhaps overoptim-istically) potential of 6 BBLS. In the DRC, fields are small: 100 MMBLS may be proven remaining, with potential at 150 MMBLS. But it has enormous frontiers – the vast Cuvette interior has seen little activity and the Great Lakes area is still untouched. In Equatorial Guinea, 2.5 BBLS has been proven already, and potential of around 3 BBLS is considered feasible. Large offshore areas are still to be leased. Gabon has experienced production decline and now 2.6 BBLS is thought to be proven remaining, with potential at perhaps 5 BBLS. Mauritania has 100+ MMBLS proven. Ultimate potential might reach another 1.5 BBLS. Nigeria's government cites 37 BBLS proven. This is expected to rise to 40 MMBLS by 2010, with potential for around another

20–40 BBLS by 2025 or beyond. This includes deepwater and ultra-deep plays, the JDZ and interior basins being opened only now. South African oil reserves are low at 40–50 MMBLS, though state agencies estimate potential at 1 BBLS (including deepwater and EEZ options expected to be available within 10 or more years). In Sudan, proven oil is over 5 BBLS and growing. Some well-informed Sudanese sources and operators estimate potential of 8–12 BBLS.

Exploration across Africa has been advancing rapidly in both producer and non-producer countries. Some of the latter are considered by the best companies now in situ to have good potential. Benin may have some 100 MMBLS. Its small area of deepwater has been little tested. As yet, Central African Republic has no commercially proven reserves, though industry sources claim potential of 1–2 BBLS. In Ghana, where there has been production and small amounts still flow, GNPC claims 800 MMBLS potential, though 50–100 MMBLS is a more likely near-term estimate (future potential is unclear). Guinea-Bissau's AGC-Dome Flore and related structures hold 1 BBLS of unexploited heavy oil in place. Kenya is now a highly targeted frontier. It has no proven reserves but is said by operators to have potential of up to 3 BBLS, a claim that awaits drilling in deepwater and ventures onshore. Liberia has offshore blocks, the potential of which some estimate at 100–200 MMBLS. Madagascar has huge quantities of heavy oil discovered and proven but yet to be commercialised. The Tsimororo HO field has P-50 reserves at 9.4 BBLS (P-90 at 2 BBLS) with recovery rates placed at 30–50%. The Bemolanga extra-heavy oil (XHO) field has P-90 reserves placed at 34.9 BBLS (P-50 at 64.9 BBLS). OMNIS suggests recoverable offshore potential at 5 BBLS, and many companies (including ExxonMobil) have now picked up acreage onshore and offshore. More leasing is expected in current and future bid rounds. Senegal has only 10 MMBLS proven and rights to an 85% share of AGC-Dome Flore waters. Much exploration is committed. Sierra Leone has awarded several blocks and potential, though still unproven, is thought reasonable. Somalia has proven oil and also has potential of perhaps 100 MMBLS in the near term and, some say, up to 2 BBLS in the long run. Though Mali has no proven oil, one informed player puts potential at 2–4 BBLS. Mozambique, an established gas producer, has deepwater plays soon to be tested. In Namibia, discoveries so far have been for gas rather than oil, but many players have now entered the onshore and offshore with oil targets. In Niger there is some 350 MMBLS proven oil, yet to be commercialised, and potential may be up to 1–2 BBLS. São Tomé and Príncipe's Obo-1 discovery indicates a large oilfield with OIIP at 1 BBLS. Talk of 4–8 BBLS potential for the entire JDZ should be treated with caution at this stage, though many do consider the area highly prospective. To date, Tanzania has been a gas play only. With Mafia Basin deepwater exploration plays at earliest stages, potential oil indicated by TPDC may total over 500 MMBLS. Significant operators such as Petrobras have deepwater acreage there. In Uganda, operators estimate potential of up to 300 MMBLS, with an initial discovery logged in 2006 with over 12 MBOPD flow combined in three horizons from one well on a field put at 30 MMBLS. A second discovery has also been made, providing new perspectives on the Rift basins and interior East Africa. Production is expected in Uganda for the near term.

There are some outsiders and frontier states with, it is believed, modest or little prospect of oil in the near term. Botswana is seeking to commercialise large coal-bed methane (CBM)-gas reserves. Eritrea has neither proven oil nor clear potential yet evident. More exploration is needed there. Likewise, Ethiopia is, so far, an exploration play, except for the large Calub gas/condensate field that might be used for LNG. There is no oil proven in The Gambia, though initial potential is put at 100 MMBLS. Guinea has no proven oil and potential is undeclared for now. No one expects oil discovery in Malawi. Rwanda has a 50% share in Lake Kivu methane gas, yet to be commercialised. Seychelles has no oil reserves but state company Seypec estimates a potential of 1–2 BBLS. Only one well has been drilled, some years ago. Togo has had no luck. Potential for its small offshore, which includes deepwater, is undeclared. Zambia is a long shot but much untested. Zimbabwe almost certainly has no hope of oil despite known and proven CBM reserves.

These estimates, though imperfect, are considerably higher and more accurate than "top-down" Peak Oil guesstimates. It is reasonably certain that proven oil in Sub-Saharan Africa currently totals around 67 BBLS. Sub-Saharan Africa's geopolitical locus and character have attributes that make it an attractive target for further exploration. For the long term, the main prize is Sub-Saharan potential oil. Estimated at up to 80–100 BBLS, this is considerable (even on a prudent discounted basis). This could amount (depending on efforts, success and discovery profile) to an annual volume of 2–4 BBLS a year in discovery potential and reserve growth for some considerable time.

In the North Africa and Maghreb countries, sizeable oil reserves exist. Algeria has reported 12 BBLS proven, with 2 BBLS in condensates; a BP estimate reckons that some 43 BBLS potential exists. Egypt has some 3.5 BBLS proven and has been mainly a gas success. Libya has over 37 BBLS proven and its state company National Oil Corporation places future potential at some 103 BBLS. Morocco currently has a minuscule reserve base of 2 MMBLS and Tunisia around 320 MMBLS proven. Thus there are in the Maghreb some 50 BBLS proven. Future North African oil potential is considered highly significant (perhaps even 100 BBLS or more), with much expected from Libya in particular. Overall, exploration in North Africa is expected to become more widespread and its intensity to increase, both onshore and in the large but still lightly explored offshore.

Africa is witness to frequent new basin openings, high discovery success, old basin maturations and second winds for older or marginal fields, an increasingly extensive and still young deepwater game, increased play testing, and widening success in the extensive and still frontier interior basins. More countries are now recorded as proven oil producers. Some traditional players will fade by 2025. There will be more reliance on EOR and marginal field deals, and more intensive exploitation of old fields to sustain volumes and credible oil status. Frontiers will be better known – and there may also be some surprises.

As Africa's oil reserves grow, so will its oil production. This is already happening. Sub-Saharan oil producers are widespread and increasing in number, as are their production zones: witness Angola, Cameroon, Chad, Congo, Côte d'Ivoire, Democratic Republic of Congo, Equatorial Guinea, Gabon, Ghana, Mauritania, Nigeria, South Africa and Sudan, with São Tomé and Príncipe and Uganda expected to join the list in the medium term. On top of that, there are the North African states to be added in.

In the near term, production of crude is likely to remain concentrated in Nigeria, Angola, Equatorial Guinea and the greater Gulf of Guinea generally, and across North Africa, with Sudan as the key producer-exporter elsewhere. In due course Uganda will develop as a Rift Valley oil producer. There are other candidates for oil production in the longer term. As petro-cartographers know well, a new African oil map is rapidly emerging, with fast-growing reserves and oil production, especially in the western Gulf of Guinea, the Maghreb, Central Africa and soon, probably, in the Eastern Margins. In future new oil producers may yet emerge, such as Benin, Central African Republic, Gambia, Guinea-Bissau, East Africa (Kenya, Tanzania, Uganda), Liberia, Mali, Madagascar, Mozambique, Niger, Sierra Leone, Somalia and Togo. Most of these countries have some assessed potential but are not quite there yet with proven oil and commercial fields for development. They can hardly be written off. Out of Africa will come news of more oil reserves and more oil production.

Peak Oil's portrayal of Africa is, of course, utterly at odds with this contemporary reality. We have estimated proven oil for the African continent at approximately 115 BBLS, with a further 95–120 BBLS future potential. In the latest Peak Oil estimates (circa 2005), however, proven oil totals 65 BBLS and new fields another 15 BBLS, making a total estimate for all future oil endowment for Africa of only 80 BBLS or so. Africa is said to have reached a discovery peak in 1961, and – despite all the evidence that production is rising – both a depletion mid-point and production peak in 2004.

The deficiencies in the Peak Oil estimates are in large part the result of the theoretical and methodological flaws that this treatise has already uncovered. They would seem to stem too, however, from a particular ignorance of Africa: its industry and its players, its particular business and investment dynamics, its history and context, and its dynamic frontier development. These deficiencies are of significant importance. Africa is the best example of a frontier continent. If Peak Oil's estimates for Africa are so extraordinarily low, what credibility attaches to its estimates for other frontiers, other continents?

Though we can hardly provide a microscopic analysis, let us at least focus on the main features of the Middle East. The region is of central importance not only because it is the most petroliferous but also because, as we have seen, Peak Oil theorists distrust its estimates (especially in Saudi Arabia). Indeed, they discount Middle East reserves by a massive 300 BBLS.

In Saudi Arabia, a $50 billion programme has recently been launched to increase capacity to 12.5 MMBOPD by 2009. This will provide a spare capacity cushion of 1.5 MMBOPD. Higher levels of production are targeted for 2013. As discussed, fears arising from the Ghawar water cut issue have been exaggerated. In the entire history of Saudi oil, only 300 NFW have been drilled. Saudi is, therefore, far from fully explored. Its potential remains significant.

If sanctions were to be lifted from Iran, its production would certainly increase (more so if there were less bureaucratic intervention). The government's aim is for further development of a few large plays and of 16 smaller fields, with a target of increasing capacity by 1 MMBOPD. Additional production is expected from, in particular, South Pars, Azadegan (at 250 MBOPD), Yadarvan and several fields in Khuzestan.

Not surprisingly, because of the security risk and the hiatus in legislation, large deals are not taking place in Iraq. There has, however, been some activity from, for example, Petrel Resources in field development and some small players in Kurdistan. Once the situation stabilises, there will be a frenzy of interest from corporate and state oil companies. While Peak Oil allows only 71 BBLS, official estimates of 115 BBLS probably underestimate ultimate reserves and recoverable potential. Before the conflict, estimates of 5–7 MMBOPD for long-term production, based on benign conditions of investment for field exploitation, redevelopments and new exploration, were not unwarranted.

In Kuwait, the North Field project development remains stalled. The much-vaunted *Petroleum Intelligence Weekly* report estimating Kuwaiti reserves at 30–40% lower than the official number of 99 BBLS was rejected by the oil minister. In the light of a major oil-gas find announced in 2006, net reserves estimates may increase. In Oman there are currently only 11 companies involved on 20 blocks. EOR investment has begun only recently. Production is projected to reach nearly 800 MBOPD by 2010. In Yemen an exploration boom is in progress. In Jordan, very much a frontier zone, the National Resources Agency (NRA) has started to see some success in acreage leasing.

Across the Middle East there is a vast area that has not yet been fully explored. Particularly important is Iraq, where almost no modern technologies have been applied, investment has been minimal, exploration tracts are large but still unleased and most blocks have so far been little evaluated. Across the entire region only about 2,000 NFW (3% of world drilling) have been drilled. In the Persian Gulf since 1995 there have been fewer than 100 NFW, 150 appraisal wells and only 5,000 development wells. This implies much room for future upstream potential.

OPEC (including, of course, members from outside the Middle East) expects its capacity to reach 37.6 MMBOPD for crude and 6.1 MMBPD for other liquids by 2010. The upstream investment plans involve more than 100 projects at a cost of around $100 billion. There is in such figures not the slightest hint of an imminent peak in Middle Eastern oil production due to reserve inadequacy.

Russia is another zone of great significance in the Peak Oil debate. The Peak Oil estimate of 90 BBLS is considerably below other estimates. Estimation of Russian reserves requires a complex understanding of reserve concepts and classifications, oil history and the players involved. An in-depth analysis by John D. Grace was published in 2005.[124] He shows that in 2003 Russian proved reserves (on an SPE basis) for existing fields stood at approximately 68 BBLS, probable reserves at 32 BBLS and possible reserves at 35 BBLS. Additionally, a low undiscovered oil estimate at 95% for conventional oil is put at some 46 BBLS (80 BBLS on the mean estimate). Leading oil reserve engineers DeGolyer and McNaughton are reported to estimate reserves at more than 150 BBLS. The technology used in Russia is not yet state-of-art everywhere and the Russian exploration play is not fully open; indeed, resource closure to foreign companies is apparent for now. Meanwhile, recovery rates in fields may be at only half of world average. Overall, Russia's resource/reserve estimates indicate a strong reserves position with high potential on the upside.

There is a discernible pattern of serious underestimation in Peak Oil numbers. In Venezuela, for instance, Peak Oil estimates 34 BBLS for reserves from known conventional oilfields and 6 BBLS for new fields, compared with the PDVSA and Oil Ministry estimate of 70 BBLS. Indeed, in 2006 Venezuela announced new oil reserves at 305 BBLS (and now claims to have the world's largest reserves), of which 235 BBLS consisted of Orinoco heavy crude potential. Though this may not be conventional oil, no one will care very much when it hits the market. Other examples of excessive pessimism in Peak Oil forecasts would not be difficult to adduce.

We should at least briefly consider the situation in the US. In Alaska the Prudhoe Bay field was shut down in 2006 as a result of pipeline problems. There has been a trend decline in production, though further exploration efforts are expected in Alaska. BP estimates remaining reserves at 17.5 BBLS with YTF at 5 BBLS. So far 15 BBLS has been produced.[125] In BP's own operations, a 1% increase in recovery efficiency would add 600 MMBLS to net reserves. BP is seeking to halve the rate of decline, thus extending production towards 2050. The costs to develop this Alaskan-wide

resource base are large, requiring significant company involvement as well as a suitable long-term tax regime.

Elsewhere new resource access and future reserve discovery
and development could be improved by a reconsideration of
open areas. At present, 85% of all American territorial waters are
excluded from the industry's access rights. The undiscovered oil
potential in the areas demarcated for possible opening offshore in
the Pacific, Atlantic and Gulf of Mexico (GOM) could allow the
tapping of up to 85 BBLS in estimated potential. In other words,
a lot of to-be-discovered oil that technically could be recoverable awaits the political passage of bills through the legislature.
An indication of the potential to be unlocked is provided by the
major hydrocarbon discovery by Chevron (Statoil and Devon) in
September 2006 in the GOM ultra-deepwater. Finds from 30,000
feet below the subsurface were announced from several fields.
The preliminary estimate is substantial and 3–15 BBOE were
indicated. Although it is too early to define actual reserves, and
they may turn out to be smaller, this indicates an undiminished
future for leased zones in the GOM, as well as signalling new
potential in worldwide ultra-deep plays.

Overall, it is clear that conventional proven oil reserves
estimates considerably exceed those used by Peak Oil in Africa,
Latin America, Russia, the Middle East and elsewhere. Undiscovered oil potential is also considered large and beyond the
levels used in Peak Oil's models. According to Abdullah Jum'ah,
president and CEO of Saudi Aramco, 18% of world crude has
been used so far. He estimates world conventional oil potential at
4.5 TBLS, a volume considered sufficient to power the globe at

current levels for 140 years.[126] This estimate of reserve potential dwarfs the numbers deployed within Peak Oil.

As we have seen, Peak Oil forecasts have relied on unnecessarily narrow definitions of reserves and potential for even conventional oil. Let us, therefore, broaden the picture by considering other kinds of oil and hydrocarbons. We should start by adding heavy crude and unconventional oil to the equation. Even with the bottlenecks now evident in oil sands projects, the production capacity for XHO from Canada and Venezuela is expected to reach approximately 5 MMBOPD by 2015. The Orinoco has enormous defined reserve potential (235 BBLS) and the current corresponding figure for Canada is 175 BBLS. The Canadian Association of Petroleum Producers considers that Canada could be producing 3.7 MMBOPD in 2015, with increasing amounts (to nearly 3 MMBOPD) from oil sands. This would make Alberta a magnet for new investors from around the world. A dozen or more companies are planning to spend $60 billion on oil sands projects over the coming decade. Other potential sources of HO-XHO include Morocco, Brazil, China, India and the US, to name a few.

Global liquids supply will also be enhanced by flows from condensates and NGL. Production of 22 MMBOPD is expected by 2015 (compared with about 14 MMBOPD now). Sources include Qatar, Norway and Saudi Arabia. Though growing, GTL production is small: by 2015 it is expected to have reached only 1 MMBPD. The higher price of crude has, however, led to new projects being examined. Natural gas plays an important role in many parts of Europe and also in the Southern Cone, Southeast Asian and East Asian LNG markets, and the fast-growing

Atlantic LNG business. Estimates of world gas reserves are not only enormous, but growing. Gas growth has had an oil back-out effect in places and is likely to do so more in the future.

Few people involved in corporate oil are under any illusion that the exploration, extraction and development of the vast world oil resource and reserve potential will be easy or low cost. ExxonMobil notes that there are "no easy answers" (though its Australian chairman has noted that a further 10% increase in recoverability would increase total recoverable significantly); Chevron has announced that the era of easy oil is over. Most oil companies have revised their strategies. Total has widened its global portfolio options and made an acquisition in the Athabasca oil sands; BP includes several energy alternatives as a very small part of its portfolio; Shell is seeking "Elephants" and "Big Cats" in portfolio realignment, as well as increasing the proportion of gas-LNG in its asset mix.

It is clear from their strategies, therefore, that companies are well aware of threats to world oil supply. This is not, however, based on any innate belief in global oil depletion. Governments are bargaining harder for assets and some have been closing doors to entry and access on reasonable terms. Large projects are also complex and take time. Conditions in the supply chain are tight in terms of engineering, geotechnical staff, drilling rigs, day rates, contractor capacities and expertise. Meanwhile, new global players, many without any industry assets, have had an impact on reserve access, exploration potential, portfolio choices, commercial interests and upstream strategies, thus adding to the difficulties and complexities of the global oil game.

19

The 21st-century
hydrocarbon game

The world oil future will not be shaped just by resources and reserves. There are many oil games on strategy and geopolitics in play. Some of them have served to reduce the access of players to reserves and forced new choices in corporate upstream options and portfolios. Geopolitical events have already had an impact, and may at times have a more profound future influence on world exploration and oil supply, irrespective of oil reserve adequacy. None of this realpolitik, however, penetrates the artificial model of Peak Oil. It would be unwise to base future portfolio and management strategy on the idea that recent times have reflected, and coming decades will reflect, a trend continuation of the past.

There are many "new players" in present and future world upstream scenarios. They affect exploration, reserve access and production profiles. They have become part of the global oil scene. Companies have in the past been accustomed to a set of commercial and non-commercial interfaces upstream that included competitors in the form of corporate oil, national oil companies and governments. Now a wider set of entities and influences affect their global portfolio and strategy decisions. These "new players" include NGOs, lobby groups, churches, "dissident" shareholders, financial institutions, ethical investment groups, indigenous communities, armed militia, terrorists, and often less organised but important "social" groups of diverse types, some hostile to foreign oil companies. This phenomenon of self-organised vested interest has arisen as the formal structures and legitimacy of international order, central authority and the nation state have been contested in the march towards globalisation. As globalisation has continued and even fractured, so such groups have proliferated. If the paradigm becomes less stable and more anarchic, it is likely that more groups may move from mere critique into overt hostility. Oil is a prime target for this global animus.

These so-called non-formal entities (some are well structured and established, so the epithet is not exact) exert a new set of influences on companies in the upstream industry. These include pressure for disengagement from contentious investments and regimes (for example, Sudan, Iran, Myanmar); campaigns for the insertion of specific non-corporate rights into the decision-making process and the ownership of oil and gas assets; requirements or pressures placed on companies to supplement investment in foreign ventures with corporate social responsibility programmes and the like; and a need felt within companies to respond to social pressures and political debate and to develop public relations strategies.

Political debates on petroleum do not take place in a vacuum.

The global petroleum industry has had to learn to participate in the global oil debate. Few oil executives, however, are well versed in managing all the political complexities. Many wish they would go away, though increasingly this is not an option. Change in world petroleum will be contentious as new oil maps emerge and as older versions of the nation state are modified. All of this will have an impact on oil players.

Petroleum is at the heart of the clash over natural resources and the power that they may bring. In some places, a ruthless struggle for resource control can be anticipated, comparable to that now found in Nigeria's Niger Delta. Nation states may well not be able to contain all future conflicts. New "iron curtains", religious walls and ethnic divides could impose heavily on global petroleum players. Corporate reserve access, development and the exploration game have already been affected.

Globalisation carries uncalculated risks – the erosion of the nation state possibly among them – leading to new divides in world society. Oil players find themselves at the centre of many of these political disputes. Powerful historical forces from the 20th century, such as communism, have left their mark on the structures, organisations and body politic of the 21st century and may blend with problematic new orthodoxies. This adds to the complexities that companies need to negotiate.

Stability, so important for the development of the global oil industry, may not be assumed. The industry's precepts and modus operandi may be contested in many areas. New geopolitical alliances could reshape world oil cartography. The ghosts of the "great game" of old may be joined by new, even less benign participants. The geological and market tapestry of the industry has already been affected by such developments. Warlords in one area and fratricidal strife in another have a direct effect on oil companies and access. The infrastructure and political architecture essential to exploration and oil extraction might become still more problematic over time.

Global oil companies will not be able to fix all these problems. Petrodollars are incapable of delivering *la dolce vita* for all, despite many companies' well-intentioned efforts to ameliorate social disadvantage in the areas where they place their assets. More companies will need to venture into the cauldron of uncertainty that is the more fragile developing world, where political earthquakes can suddenly occur. Much of the terrain where companies must invest might prove highly sensitive in terms of institutional order. Companies confront, on top of those issues familiar to the oil industry, many new threats: dissidents, sects, opposition parties, ethnic rivals, ideologues, activists, armed militia and mendacious politicians, whether aligned or not with the governing elites with whom corporate oil must seek to do business. Deals struck with foreign states and corporate elites may not always be legitimised, sustained, or uncontested in the investor's home base.

A fragmented, even Balkanised world brings greater contestation of legitimacy. The ethical standards and morality required for transactions to be made may not always be on offer in some parts of the newly emerging petroleum world. Self-interest may increasingly conflict with moral absolutism in more turbulent times and in difficult regimes. This could narrow the boundaries for acceptable investment choices and compromise. Nobody knows how many political convulsions there may be in regions such as the Middle East, Central Asia, Africa and Latin America over the next quarter of a century. The "Heart of Darkness" was terra incognita, and Conrad's paradigm is not restricted to the African situation: unknowns still abound in the futures of many oil-reserve-rich states (for example, Iraq, Iran, Russia, Venezuela and even Indonesia).

Debate on petroleum, the most "political" of commodities, in forums outside the industry (formal, informal and in cyberspace) is likely to sharpen. Corporate oil is required to engage with issues arising both within and beyond the industry. The contentious issues include: human rights and oil rights, both vexed questions in a turbulent, oil-driven world; armed security in oilfields; rogue states and oil reserve access; corporate oil's presence in conflict zones (actual or potential); oil sanctions and related restrictions on portfolio access and choice; groups such as Transparency International and Global Witness seeking to modify traditional oil industry practices; indigenous peoples' rights and retrospective oil claims; the so-called oil curse found inside the oil and poverty nexus, for which corporate oil is increasingly blamed; a swathe of Kyoto-type issues associated with the world oil industry; pressures from the great and the good seeking to make corporate oil conform to the latest political fashion; and a mixture of issues in the developing world, both global and local, which attract a huge number of NGO interests (oil, it seems, is their new "business of the day"). There are also environmental issues concerning shipping, oil spills, effluents, fishing, rainforests, conservation, animals, endangered species, threatened ecosystems and biodiversity.

Each such issue attracts NGOs, networks, informal lobbies, organised protest, social comment, editorials, direct action and demands for stricter policies, sometimes backed by local or international law, "soft" law, or state mandate. Whatever the future of the oil industry, it is unlikely to follow a smooth path – unlike, of course, those statistically adjusted curves to be found in Peak Oil forecasts. Though the constrained above-ground oil world now emerging will not change the oil resource endowment per se, it could have an impact in terms of accessible commercial reserves. Company strategies and portfolios have already begun to adjust.

From the scramble for hydrocarbon resources, new geopolitical strategies are emerging, designed to alleviate threats to stability. This has led to the attempt to establish various "empires of oil", so to speak, among them American, Russian, Chinese, Middle Eastern, Indian and Latin American (the regional alliances built by Chavez). Africa's largely fragmented character generally precludes this model except on a limited state basis. Yet overall, a new global paradigm is emerging, though it is one that may shift again if the price of crude falls to a lower range for an extended period, and if significant new spare capacity is installed, changing the structural relationship between hydrocarbon geopolitics and oil markets and reserves.

Oil markets cannot, of course, resolve all the struggles, conflicts and disputes that the oil industry faces. Government is inevitably prominent and will make all manner of interventions – not simply in periodic military action, but also *inter alia* through global and regional diplomacy, soft loans and tied aid, strategic bilateralism, intergovernmental deals, state energy investments, debt write-offs, collusion and the formation of quasi-cartels. Against a backdrop of fractured globalisation, these developments make for more complexity and a tougher interpretation of the driving forces shaping the future upstream industry. They may also lead to images of a new oil-driven medievalism in which mercantilist interests sublimate foreign policies as oil moves higher up the global agenda. With so much of the developing world in a state of flux, alliances can be fragile and subject to new controls exercised over crude reserves and opportunities.

This context provides fertile ground for the global enemies of corporate oil. The oil game is very different in nature from that of a couple of decades ago. It is well known, for example, that terrorist groups such as Al-Qaeda have an oil-focused agenda. So do many NGOs worldwide. *De jure* sanctions and *de facto* boycotts restrict reserve development in significant places. These paradigm changes have an impact not only on corporate oil, but also on national oil companies (as has been seen already with, for example, Petrobras in Bolivia).

Yet there remain undiscovered oil worlds in a relatively undiscovered future world in general. Each year and each decade reveals more about the reserve potential and development options for the industry. There are still many places for it to travel to and explore, even if allowance for future surprise and, indeed, error needs to be made. In such a world, Peak Oil's grand mythology that "all is known" looks distinctly out of place.

Part 5

TWILIGHT IN THE MIND

This inquiry has travelled a long path. It has sought to understand the context within which the Peak Oil movement has developed, the reasons for its existence, and for the existential angst related to it. It has examined the theory and evidence for Peak Oil on its own terms and from the wider perspectives of economics, strategy and geopolitics. The aim has been to assess the validity of Peak Oil's case, focusing on the rationale for its exclusive assumptions, the geologic presumptions on the supposedly "known world", the selection and discounting of data, shifts in the forecasts themselves and projected outcomes. It has also reviewed some of the conceptual and empirical criticisms that Peak Oil theory has attracted.

Who can be the best judge of this debate? The Australian government set itself that task by deciding to hear the evidence, as it were, in a debate organised in 2005.[127] Bakhtiari ("the Prophet Ali"), among others, was invited to present the Peak Oil case at one of the forums on which the official report was based. In the event, perhaps not surprisingly, the bureaucrats danced around the hot political fires in classic hedging fashion.

A report was duly issued that at least highlighted some pertinent issues. Attention was given to the inaccuracy of Peak Oil forecasts – even Geoscience Australia was found to have exhibited "forecaster's droop" (the propensity for consistent pessimism and error). Peak Oil's backdating method was noted as a form of historical revisionism (that is, highly dubious). There was much discussion of conflicting URR estimates. More radical was the consideration given to the use of changing URR/P ratios (in place of R/P). This was perhaps a step forward, suggesting a more dynamic formulation of the problems (one hardly compatible with Peak Oil paradigms). No verdict, however, was announced – testimony perhaps to the Australian sense of fair play – and wide gulfs in the debate remained.

We can hardly leave it there. We need to arrive at some judgement of the issue. Perhaps at this point we should borrow from the Renaissance the services of Niccolò Machiavelli, that renowned arbiter of court intrigues concerning men of ideas and of power, their views on strategy and statecraft and the world at large at the time.

20

Machiavelli's judgement

We are faced with either the reality or an image of twilight. The question Machiavelli would ask is whether the much-forecast "twilight in world oil" is manifesting itself in the world or whether it is a mirage formed by "twilight in the mind". As a realist par excellence, Machiavelli would want to observe the world around us, including all the multiple events that conspire to shape the global oil game. His judgement would depend not on what the sacred texts have prophesied, but on whether or not the Holy Grail has been found. He would examine all the concepts, theories, texts and subtexts, and the minds from which they have arisen in just that spirit – that is, with ruthless objectivity.

Peak Oil is in some ways comparable to a form of global central planning. There is a model to which all data and circumstance must conform. Watching it in action, it is easy to become preoccupied with the numbers: changes in the data do, after all, turn out to have serious implications. But it is important not to take one's eye off the underlying assumptions, however well concealed they may be. Amnesia about its assumptions is like ignorance of history: it condemns us to repeating errors.

The inner linearity of the model reflects the parametric fixity of Peak Oil theory. The structure of this model is rigid. There is no feedback loop and only minimal interaction (at best) between man and nature. Nothing can alter the unalterable. We find in this mindset only steady-state concepts of stasis. Humanity is condemned to march in one direction, towards the depletion of oil, to the monotonous beat of the model's equations.

The phenomena of the oil world – the dramatic stop-go cycles of discovery and production, the unpredictable volatility of the market – are anything but smooth. It is a rough, jagged, multi-dimensional world. It is difficult to imagine anything less like the seductive, two-dimensional smoothness of the Peak Oil model, which itself follows seamlessly from the prejudgements on which it is based.

The conception of the past as an analogue for the future (written into Peak Oil's projection ethic) leaves no room for any sense of paradigm shift. It depends on the technique of backdating reserves and the minimisation of the undiscovered with a diminished view on the future potential of technology. It offers the dubious pleasure of time-warp nostalgia and revisionist explanations of continual failures to peak. The equation of equivalent behaviour across time (with regard to crude price, for example) assumes that time/periodicity is constant and seamless. The history of the future, as it were, must be neglected like the history of the past.

The problem of *ceteris paribus* assumptions, of course, is that other things are rarely equal for very long. Such a belief can be sustained by studiously ignoring the messy, inelegant, non-parametric, above-the-ground world of economics, technology, management, company strategy, government policies and geopolitics. An ostrich burying its head in the sand has to treat a lot of variables as exogenous. To look only underground – at some geologic control based on the notion of a *quantum fixe* – involves a comparable set of assumptions. It is the equivalent of asking the world to stop so that you can get off.

This treatise has at various times alluded to the quasi-religious character of the Peak Oil movement. Certainly concepts of the "all known", the "inevitable" and the "immutable" smack more of theology than of natural philosophy. Resting on such conceptions, the debate takes on the form of a semi-religious contest for ascendance and search for legitimacy. It manifests as a power game over who controls access to the knowledge found in the Almighty or holds the hand of God.

From this perspective, the view of the oil frontier as an important, dynamic phenomenon is a heresy. When reserve growth is denied or airbrushed into backdated discovery history, this imputes a zero value to the process in future. Yet most companies' operational strategy depends partly on this factor. In some cases, whole portfolios have been built on it. Attributing such importance to the not yet known would, however, cause a quasi-theological difficulty for Peak Oil theory, equivalent to asking the Vatican to tweak and adjust the Ten Commandments. In its view of the frontier, Peak Oil theory will allow only the odd skylight to be opened here and there. It is not enough to shed daylight onto the dark world of oil.

It is as if the Peak Oil movement is playing mah-jong while the oil world is getting on with its game of chess. The former, though not without an interest of its own, cannot accommodate the moves and tactics of the latter. In just such a way the Peak Oil game fails to accommodate such dynamic forces as the changing price of crude and its effects (on substitution, for example, and even on reserve growth), or the formation of company strategy or statecraft, or the development of technology.

Ironically, given its disdain for economics, Peak Oil theory is reminiscent in particular of the neoclassical paradigm within that discipline. Order above disorder is the leitmotif of both: there is no room for turbulence, or random items, or long-run disequilibria, or chaos. The equilibrium in Peak Oil theory is, of course, depletion. Yet as forecast succeeds forecast and peaks shift continually, equilibrium seems always to be postponed. We are left with a bizarre state of stable instability – but with no awareness that it is time to change paradigm.

Peak Oil presents a harsh view on the intractability of the Earth and perhaps, even, on the meanness of God: there is only so much oil – and damnation thereafter. In this sense Peak Oil theory, for all its finite assumptions, claims to know the infinite. Yet we have only to admit the concept of oil as a *quantum flux* to realise that this is untrue: Peak Oil geology does not know all that is in the (non-linear, dynamic) disposition of the Heavens.

A key element of Peak Oil cosmology is the conception of time. Time in Peak Oil theory adopts linear shape and neutral form. The theory itself seems to be constructed like some Newtonian clockwork mechanism, built to measure in a relentlessly unidirectional manner without need of, or scope for, ingenuity or imagination. Oil history, however, seems far from linear and can hardly be assumed so for the future. In particular, time is never neutral in the realm of knowledge, whether that knowledge is of geology below the ground or the world oil business above it. Our technical knowledge over the coming decades will not stand still, so why discount its advance?

Peak Oil is above all about the forecast. Yet forecasting the future has perhaps become even more difficult than it used to be: it seems there are not enough crystal balls of quality to employ. Whatever is the case with nostalgia, the oil future is not what it was. A complex paradigm is in play, one that will probably change with the passage of time. We need to treat the data as subjective or imperfect and our judgements as uncertain. Peak Oil's sense of certitude stands wholly against such flexibility. Of course, the earth used to be flat, and the sun once went around the earth, and there was a time when there was no oil in Saudi Arabia.

A dynamic world of oil undoubtedly exists, some of it indeterminable. This is the way the world works, and history too, rather like the paradox of the chicken and the egg. Peak Oil theory, however, points for ever to one image, that of the Apocalypse. It offers us "Apocalypse always" embedded in a tradition of repeated failure. The essentially non-linear world oil industry has refused to conform. The model pays a heavy price for having built its foundations on excessive immutability and insufficient understanding of the deep complexities within and around the world oil game.

It is best to know best what we know not – then we can add missing links over time to approximate the realities of world oil and tread warily into the long run (a destination at which we never arrive). We need to retain the widest geological imagination and allow for both the expectations of experience and the unexpected. No model can "model" the upstream world without permitting human ingenuity to exist in all facets of the industry and oil-energy markets. Peak Oil needs to build better global, corporate and geopolitical intelligence on the wide influences that shape the world oil future. It needs to make allowance for unknowns and grapple with the twin beasts of risk and uncertainty.

The Peak Oil paradigm also lacks a sufficient appreciation of economics. Traditional economists using *ceteris paribus* allow "everything to be permitted, except those things that are forbidden". Its inverse variant (after all, economics allows for many models) is where "everything is forbidden, except those things that are permitted". Peak Oil seems to be about "everything being forbidden, including those things which are permitted". What should be sought is the non-linear in a model "where everything is permitted, including those things that are forbidden". This is a world of unbounded definition, one that is closer to our world oil realities.

This treatise has moved both within and beyond Peak Oil theory and has covered a wider canvas as it has done so. Peak Oil theory is found at one end of an extreme edge in the theoretical spectrum. There are many rich choices in the world oil future that confronts mankind. Much remains a matter of dispute. Various conventional wisdoms on the future of oil continue to compete with each other for ascendancy. Though no one model appears unimpeachable, theory and evidence, Machiavelli tells us, lean strongly against Peak Oil.

2 I

Sunrise in the Kalahari sands

A narrow knowledge base built on the purely technical would not have enabled the Khoi-San to survive within the harsh confines of the Kalahari. Theirs was a world "made up" with rich imagination and intuition concerning all the elements, those found on, below and above the ground on which they trekked for over 20,000 years. Not for them restricted vision or constrained empiricism or presumption that each sunset will be the last: the secret of their survival was adaptation.

It is a paradox of history that the long-neglected Kalahari sands are now reported to conceal some unexpected energy resources, and so our long trek must return to the desert where a new dawn is born beneath African skies. The Kalahari Gas Corporation is developing CBM-gas reserves, estimated at around 12.5 TCF in potential, to provide for Botswana's growing power-energy needs, export to nearby markets and substitute for existing fossil fuel usage. While by no means wholly offsetting the lack of oil reserves in Botswana, this micro-case of local energy adaptation is one of many the world is witnessing and can expect in future.

Perhaps Peak Oil and even its critics should go on safari and look at the Okavango Delta, a lacustrine system in the middle of Central Africa on the edges of the Kalahari, into whose sands the Okavango's waters drain. This was once home to the Khoi-San and still is to many modern-adapted Bushmen. It is a world where the power of adaptation is readily witnessed. Adaptation here is needed for survival.

The Okavango once "seen" as its inhabitants have seen it exhibits natural non-linearity and complexity. There the mind will find curious and beautiful self-replicating shapes in which many parts mimic the whole. Unique features analogue and hold lessons for competition landscapes and competitors as found in the flora and fauna located there. The flora establishes the environment within which the animals must compete and find survival. The delta provides one of the path-breaking paradigms of nature, a unique natural reality offering brilliant images of rare beauty, harmony and order amid continuous turbulence.

As you move into the world's largest inland delta, multiple images imprint deep impressions. There you find a splendid panorama of diverse shapes and repetitive patterns etched in the earth's rich surface and magnificent uncontaminated skies. The "sets" of waterholes, meandering river courses and spotted dry pans (awaiting the episodic but annual flood stream) replicate their commonality and diversity across the horizon. This is one of Africa's deepest recesses and the heart of a natural and enduring miracle: the origin of man.

Encountered further inland towards the heart of the delta are denser waterways and richly knitted landscapes. The tapestry shifts perceptibly as new boundaries are crossed. One thing leads to another. The land of a dry sunburnt plain blends in harmony and contrast with complex but changing river channels. Red colorations merge with the verdant foliage, lapped by the pristine blue waters of a vast array of arterial passages crafted by connecting links in fragile but integrated ecosystems inhabited by stunning mammalian populations.

Patterns in the bush capture attention. Flora and fauna mix in a wild, turbulent and occasionally contradictory manner. This vast animal kingdom boasts over 1 million predators, grazers, carnivores, mammals, birds and assorted innocents, some of which feature high on the weekly à la carte menus of lions, leopards, hyenas, cheetahs, wild dogs and other meat-eaters. It is a world of diverse species, including the majestic "big 5" (elephant, lion, buffalo, leopard and rhino), that roam its territory, at the top of the food chain. The nature, patterns and selection of predation illustrate the issues of social survival and constrained success.

This world of "discovered chaos" holds immense uncertainty, fragility, turbulent order, regular surprise, sudden jagged edges, deep patterns and unbridled continuity, and shows itself in the Okavango's multiple faces and facets. The delta's millennial mysteries present diverse inhabitants – animals and man – with the risks inherent in life and survival. These uncertainties are beyond normal control and outside the boundaries of the easily "managed options". Here, for all living entities, dawn to dusk is an eternal question of survival. While twilight is real it is superseded by an inevitable dawn, manifest as both raw and primitive.

Coherent and temporary balances are periodically threatened by new interventions and occasional but inevitable acts of

God. One day the rains come, or the floods exceed expectation, or drought strikes a corner of this immense habitat, or rampant fire sweeps away an entire savannah bush land system that has until then successfully stood the test of time and the ravages of unrelenting nature. Fragile stability is inherent in the Okavango's interrelated and highly sensitive dependencies of different fauna, each subtly touched by the other, and myriad unpredictable externalities. The shifting elements of light and water conspire and combine in an apparent dynamic hierarchy of continuously changing conditions to offer survival and success or catastrophe to the Okavango's constituents. There is no escape from the natural order and the built-in non-linearity found at work in the Okavango. And as it is in the delta, so it is also in the world.

The raw and periodic turbulence of explosive life forces – in grass and trees, blood-red sunsets, electric dawn surprises, fire and rain – shatter the calm of illusory tranquillity and apparent certainty perceived by the conscious mind. Not all is known. There are many elements of the inexplicable and mysterious to be sensed and imagined. The subconscious and sometimes hidden implosive patterns are endless: in the rich foliage shapes, differential herd configurations and synchronous moments of cloud and atmosphere, and in the silence the ear finds interposed between unstoppable sounds which form a continuous sequence in the Okavango's history.

The delta's history stretches back to the dawn of time, to man's first steps. Each moment is a static temporal frame. A ceaseless resonance intimately connects each fragment and separate decibel to a mega-billion others throughout the world. The tracker listens and intuits all this magnificence in order to "read" the next primal act in the Okavango's future. Therein are contained the secrets embedded inside the Okavango's continuity. Its inhabitants, both animals and Bushmen, must learn many secrets just to survive, for to survive here is to at least succeed within an unforgiving and resource-constrained environment.

This natural world highlights many lessons for man, who must coexist alongside numerous changing landscapes or paradigms found in changing shapes and sometimes not readily discernible patterns. In this original world, barely touched and pristine, there exist signs and secrets about the tracker that can and should be learned: as analogue, template and strategy. They can be used as a basis for concept, thinking and continued existence. Any safari is a journey intended to reveal insight, break the chains of parametric imprisonment within our restricted images of the world, and transform any twilight in the mind.

Notes

1   See Notes on further reading on page 221.

2   The idea of *quantum fixe* is used here in opposition to that of *quantum flux*. The former denotes a finite amount of oil, already determined; the latter denotes a volume of oil that varies with conditions above and below ground.

3   C.J. Campbell, *Oil Crisis*, Preface, Multi-Science Publishing Co. Ltd, 2005. This is the leading text in Peak Oil and an update of C.J. Campbell, *The Coming Oil Crisis*, Multi-Science Publishing Co. Ltd, 1997. For detail on this author's writings, see the extensive bibliography found in *Oil Crisis*.

4   Marshall Sahlins, *Stone Age Economics*, Tavistock Publications, 1974.

5   Louis Liebenberg, *The Art of Tracking: The Origin of Science*, David Philip, 1990.

6   See James Gleick, *Chaos*, William Heinemann, 1988, a seminal text that brought many of these notions to the world stage. Also Benoit B. Mandelbrot and Richard L. Hudson, *The (Mis)Behaviour of Markets*, Profile Books, 2004, for some ideas and concepts of one of chaology's leading thinkers.

7   See Paul Roberts, *The End of Oil*, Bloomsbury, 2004; Jeremy Leggett, *Half Gone*, Portobello Books, 2005;

Andrew McKillop and Sheila Newman, *The Final Energy Crisis*, Pluto Press, 2005.

8   The best guide to Hubbert's theory and extensive work done on the United States, later with application to the world, may be found in Kenneth S. Deffeyes, *Hubbert's Peak*, Princeton University Press, 2001. See also Kenneth S. Deffeyes, *Beyond Oil: The View From Hubbert's Peak*, Hill and Wang, 2005.

9   See Campbell, op. cit., for the range of works written on Peak Oil.

10   Here I deploy this concept from sociology to refer to a relatively small group of outsiders, recent in their origins, separated from the mainstream at large, cohesive in structure, with a system of core notions or beliefs (in this case non-religious) that are considered novel, and are seen as in some opposition to conventional ideas.

11   Campbell, op. cit., p. 109.

12   See C.J. Campbell and J.H. Laherrère, "The world's supply of oil 1930–2050", Petroconsultants, Geneva, 1995.

13   Campbell, op. cit., p. 195.

14   Ibid., p. 202.

15   See www.peakoil.net for ASPO-5 Conference 2006, 18-19 July, Pisa, Italy.

16   Ibid.

17   The parabolic fractal is a distribution or logarithm of the frequency or size of entities in a population, as a quadratic polynomial of the logarithm of the rank, which

may markedly improve the "fit" over a simple power–law relationship.

18 See notably Louis Arnoux, "27 Years After Peaking, Where To?" (text distributed at ASPO-5, Pisa, July 2006).

19 ASPO-5, op. cit. See Luigi De Marchi, "The psychopolitical roots of the energy crisis", Pisa, 2006.

20 See Renato Gusso, "World Oil Depletion: Diffusion Models, Price Effects, Strategic and Technological Interventions", ASPO-5, Abstracts List, op. cit.

21 ASPO-5, op. cit. See Mamdouh G. Salameh, "Peak Oil and the Rising Crude Prices", Pisa, 2006.

22 ASPO-5, op. cit. Here see Andrew McKillop, "The impossibility of 'market solutions' to Peak Oil, the emerging financial, monetary and economic crisis, and Energy Transition", Pisa, 2006.

23 Michael Meacher, "Our Only Hope Lies in Forging a New Energy World Order", *The Daily Telegraph*, 26 June 2006.

24 Sabina Morandi, *Petrolio in Paradiso*, Ponte Alle Grazie, Milan, 2005.

25 ASPO-5, op. cit. See Robert Hirsch, "Mitigation of Peak Oil – More Numbers", Pisa 2006 (no text provided). Also see Robert I. Hirsch, *Peaking of World Oil Production: Impacts, Mitigation & Risk Management*, SAIC, February, 2005. See also Hirsch's testimony at the Peak Oil Hearings of the US House of Representatives, 7 December 2005, a record of which may be found at www.globalpublicmedia.com.

26 Bryant Urstadt, "Imagine There's No Oil: Scenes from a Liberal Apocalypse", *Harper's Magazine*, August 2006.

27 See Nostradamus Online at www.nostradamusonline.com, in particular Chapter 4 (The Time of Troubles).

28 Arnoux, op. cit.

29 Ibid.

30 Discussed in Life After the Oil Crash website (www. lifeaftertheoilcrash.net), ibid., referring to a Deutsche Bank Report.

31 See Richard C. Duncan, "The Peak of World Oil Production and the Road to the Olduvai Gorge", November 2000; text found in www.hubbertpeak.com.

32 The internet is rich in Peak Oil content, especially on the social model and its dramas. See, among others, www. lifeaftertheoilcrash.net and www.fromthewilderness.com.

33 James Howard Kunstler, *The Long Emergency: Surviving the Converging Catastrophes of the Twenty-First Century*, Atlantic Books, 2005.

34 Roberts, op. cit.

35 Michael C. Ruppert, "The Paradigm is the Enemy", found in www.fromthewilderness.com where this and several similar texts may be read. Ruppert seems to be intent on inducing the Peak Oil community to go beyond mere projection and disengage both as individuals and as a group from the "global economic paradigm". Many seem to him reluctant to do so. While the poor may always be with us, few of the rich wish to disengage from wealth and hand it over to the impoverished. And again, many

Muslims are clearly devoted but few wish to engage in jihad on the military front. This appears as the old dilemma between theory and action. Lenin and the Bolsheviks understood this matter well.

36 See Michael Fumento, "Doomsdayer Paul Ehrlich Strikes Out Again", *Investor's Business Daily*, 16 December, 1997; and also on www.junkscience.com.

37 George Monbiot, "Crying Sheep", 27 December, 2005, posted on www.monbiot.com. Here we are informed also that after reading 4,000 pages of reports on world oil, "I know less than before I started", except that "the people sitting on the world's reserves are liars". It's rather lean intellectual pickings for so much work and effort, one has to conclude.

38 Bjorn Lomborg, *The Skeptical Environmentalist*, Cambridge University Press, 2001.

39 Richard Heinberg, *The Party's Over: Oil, War and the Fate of Industrial Societies*, New Society Publishers, 2003.

40 Richard Heinberg, "Smoking Gun: The CIA's Interest in Peak Oil", 15 August 2003, and printed in www.fromthewilderness.com.

41 See www.globalpublicmedia.com for testimony transcripts.

42 More from Bartlett is found in "Peak Oil is Not a Theory", 8 December 2005, where commentary is given on the testimony – see www.fromthewilderness.com. See also an article by Tom Whipple, "The Peak Oil Crisis", *Falls Church News*, 10–16 November 2005.

43 Peter Tertzakian, *A Thousand Barrels a Second*, McGraw-Hill, 2006.

44 McKillop, op. cit.

45 For reportage and comments on ASPO-1 to ASPO-4, see various articles by Michael C. Ruppert in 2003, 2004, 2005 found on www.fromthewilderness.com.

46 Julian Darley, *High Noon for Natural Gas*, Chelsea Green Publishing Company, 2004.

47 Matthew R. Simmons, *Twilight in the Desert*, John Wiley & Sons, 2005, discusses this issue in relation to Saudi oil reserves and the world impact.

48 Thomas Kuhn, *The Structure of Scientific Revolutions*, University of Chicago Press, 1970.

49 For an excellent assessment of the origins and leading thinkers in economics, see David Warsh, *Knowledge and the Wealth of Nations*, W.W. Norton & Company, 2006.

50 The original work was done in tandem with Campbell. See Campbell and Laherrère, op. cit. But there were many antecedent publications and reports independently written, and many more subsequent. For a fuller list of these writings, see Campbell, op. cit. Numerous articles written by Laherrère may be found in several websites, including in the presentations made at the ASPO conference.

51 McKillop and Newman, op. cit, p.133.

52 The Cairn/Shell deal in the Rajasthan oil play in India post-2001 is a pertinent example on different technical views taken on basin prospectivity and potential.

53 See documents on resources/reserves and reserve valuation
that can be found on the website of the Society of
Petroleum Engineers, notably under Petroleum Resources
Classification System and Definition, including an SPE
schema shown in Appendix 2.

54 One could execute a double empirical modification to the
model – changing the numeric of both measured stock
(reserves) and flows (oil supply-demand). But this would
imply evidence of the oil demand growth rate softening
or even "demand destruction" (if, say, crude prices rose
substantially on historic trend), maybe oil substitution
(from gas and/or other fuels) and/or other deeper
behavioural changes in the markets (such as accentuated
oil conservation). It would allow the economic into the
model.

55 Available from the SPE website (see also Appendix 2).

56 See Ian Cummins and John Beasant, *Shell Shock*,
Mainstream Publishing Company, 2005, for an extensive
review of the Shell saga and related reserves issues.

57 SPE, Petroleum Resources Classification System and
Definition, op. cit.

58 Ibid.

59 Ibid.

60 There are many classic texts on strategy: but see an
excellent overview in Robert Greene, *The 33 Strategies of
War*, Profile Books, 2006.

61 See Campbell, op. cit., p. 135.

62 For key texts and works by these different authors, see
Peter R. Odell, *Why Carbon Fuels Will Dominate the
21st Century's Global Energy Economy*, Multi-Science
Publishing Company, 2004 and Campbell, op. cit.

63 See Campbell, op. cit., Chapter 10 (Economists Never Get
It Right), where the upstream risk issues are discussed and
minimised.

64 These are statements from Campbell, op. cit.

65 Warsh, op. cit., provides a thorough insight on the
contributions and ideas of many of these leading thinkers
in economics, including John Maynard Keynes, Thomas
Schelling, Joan Robinson, Charles Cobb, Paul Douglas,
Roy Harrod, Milton Friedman, W. Arthur Lewis, John
Kenneth Galbraith, Paul Samuelson, Kenneth Arrow,
John Nash, James Meade, Gary Becker, Wassily Leontief,
Joseph Schumpeter, Robert Solow, John von Neumann, Jan
Tinbergen, Franco Modigliani, Jadgish Bhagwati, Oscar
Lange, Amartya Sen, Nicholas Kaldor, Gunnar Myrdal,
Ragnar Firsch, John Hicks, Paul Romer. Other specific
analysis and work on oil and economics has come from a
wide range of economists over the years, notably Maurice
Adelman and Peter R. Odell among others.

66 While there is a voluminous literature on John Maynard
Keynes, for a rounded view on Keynes and his work in the
context of the times, see Warsh, op. cit., Chapter 8 (The
Keynesian Revolution and the Modern Movement).

67 In an aside and anecdote, it was also from economics
that William Nordhaus derived a long historical index

of lighting – the output derived from fuel, or oil and energy largely – to illustrate the power of technical change in human energy use: this flowing from measures around those not based on the input cost of crude but the output-uses to which it might be put. In this case it was exemplified by the cost of lighting a room at night expressed in cents per lumen hour. This showed that for almost 40 centuries there was barely a change in this cost, and suddenly around 1800 the cost of light fell precipitously, as if off a cliff. This is just one example. Peak Oil might riposte here to Nordhaus that this advancement is precisely what would be at risk in oil depletion – but then only if they are right in degree and timing, and the human ingenuity that has intervened with success across many centuries failed the world in some unique and untimely manner. This is another lesson in dynamic adaptation and technical-economic change and the care that should be taken not to neglect its power to affect the future world oil game. See details in Warsh, op. cit., Chapter 24 (A Short History of the Cost of Lighting).

68 See Warsh, op. cit. About 20,000 economists are members of the American Economic Association, and an equal number practise as non-members. This does not include the non-American world. Perhaps around 7,000 people are members of the International Association of Energy Economists, with as many more in the energy business who are not. The economics profession is also considered a relatively tough one, with around 2,000 PhD graduates

worldwide each year compared with the minting of 7,000 engineers, 15,000 physicians, 40,000 lawyers and 120,000 MBA graduates annually in the United States alone.

69 For an overall world economic analysis of oil and energy for the future, probably the most comprehensive in the field of contemporary economics, see Odell, op. cit. Odell provides an extensive bibliography not only on his own prolific writings over the past 40 or so years on the oil industry, but also those of many other economists.

70 There are numerous texts and books, but see Richard Steinmetz (ed.), *The Business of Petroleum Exploration*, AAPG, 1992, for some basic concepts.

71 Citations are taken from Deffeyes, *Hubbert's Peak*, op. cit.

72 See variously, C.J. Campbell and J.H. Laherrère, "The End of Cheap Oil", *Scientific American*, March, 1998; C.B. Hatfield, "Oil Back On The Global Agenda", *Nature*, 387, 1997; R.A. Kerr, "The Next Oil Crisis Looms Large – and Perhaps Close", *Science*, 281, 1998.

73 For a list of peak dates given since 1972–2004, see Bernard E. Munk, "The End of Cheap Oil, Once Again: Geopolitics or Global Economics?", 10 October 2005, Foreign Policy Research Institute (see www.fpri.org).

74 Ibid. and Munk, op. cit., provide a range of estimates done by various parties at different times.

75 See texts by Freddy Hutter of Trendlines on www. trendlines.ca. Hutter acts as a watchdog on ASPO and other Peak oil projections and has noted many changes in estimates over the years.

76 Odell, op. cit., who also provides information on historical forecasts for Peak Oil going back several decades.

77 Daniel Yergin, *The Prize: The Epic Quest For Oil, Money and Power*, Simon & Schuster, 1991.

78 For a view on all this conflict, see Michael T. Klare, *Resource Wars*, Metropolitan Books, 2001.

79 See CERA, "Oil & Liquids Capacity To Outstrip Demand Until At Least 2010", 21 June 2005, Press Release, and CERA, "World Oil Capacity Up To 25% By 2015", 7 December 2005, Press Release.

80 The idea of "spy companies" is one much used and enjoyed by Laherrère. Of course, espionage is an old art and has been around the oil game since inception.

81 Leo Africanus, named so by the Medici Pope Leo X, wrote what are now acknowledged as the first accounts of African exploration in *Descrittione dell'Africa* (The History and Description of Africa), published in 1550. Africanus provided much exploration knowledge before western explorers set foot inside Africa.

82 Petroconsultants, *Dedicated to Harry Wassall 1921–1995*, 1996.

83 See here the testimony of Robert Esser at the US House of Congress Hearings, op. cit.

84 Simmons, op. cit.

85 It is one of the curiosities found within and around Peak Oil that unusual notions on data are so liberally entertained. So we read all this speculation about Petrologistics from Simmons, as well as claims that the

Saudi reserves are just not what they report; and from inside Peak Oil we hear that most government numbers are flawed, dodgy, misleading, based on errors and the like. We are invited to summarily cut off anywhere up to 300 BBLS from Middle Eastern oil reserves, for instance, and to accept significant discounts on many other countries' reserves – all on the basis of Peak Oil's assumed authority and superior knowledge on world oilfields. On some occasions we need to suspend our belief and on others just laugh.

86   Chinese industry sources indicate much interest in this area, and have advanced sizeable estimates of oil (and gas) potential. But the lack of exploration makes such claims still unproven.

87   Deffeyes, op. cit., p. 67.

88   Ibid., p. 71.The reference is to Adelman and Lynch. Even so, economics is a lot more nuanced than this statement suggests, although without investment there is likely to be little exploration and less development.

89   Discussed extensively in Deffeyes, op. cit. See G.K. Zipf, *Human Behavior and the Principle of Least Effort*, Addison-Wesley, 1949.

90   Deffeyes, op. cit., p. 127.

91   Albert A. Bartlett, "An Analysis of US and World Oil Production Patterns Using Hubbert-Style Curves", *Mathematical Geology*, Vol. 32, No. 1, 2000.

92   J.D. Edwards, "Crude oil and alternate energy production forecasts for the twenty-first century: the end of the

hydrocarbon era", *American Association of Petroleum Geologists Bulletin*, Vol. 81, No. 8, 1997.

93 Deffeyes, op. cit., p. 149.

94 Ibid., p. 157.

95 Ibid., p. 158.

96 Campbell, *Oil Crisis*, op. cit., p. 319. In general, what we are presented with here is an end of geology and end of exploration syndrome.

97 C.J. Campbell and J.H. Laherrère, *Scientific American*, op. cit.

98 C.J. Campbell, *Oil Depletion – Updated Through 2001* (see www.oilcrisis.com).

99 See www.fromthewilderness.com, "Interview with Colin Campbell On Oil", by Michael C. Ruppert, 2002.

100 In McKillop, op. cit., Chapter 2, p. 41.

101 Many texts by Laherrère may be found on www.oilcrisis. com including: "Upstream potential of the Middle East in the world context", May 1996; "Distribution and Evolution of 'Recovery Factor'", Paris, 11 November 1997; "Reserve Growth: Technological Progress, or Bad Reporting, and Bad Arithmetic?" (from *Geopolitics of Energy*, Issue 22, No. 4, April 1999). Also "Is USGS 2000 Assessment Reliable?", World Energy Council, 19 May 2000. See also "Peak Oil and other peaks", CERN, 3 October 2005, and "When will oil production decline significantly?", European Geosciences Union, General Assembly, Vienna, 2 April, 2006. See also Letter to Phil Watts, 13 October 2001, found in www.oilcrisis.com. The comments here

on Laherrère are drawn from these texts and from the
recent ASPO-5 Presentation 2006, similar to the Vienna
Presentation.

102 See peakoildebunked.blogspot.com, 283: "Laherrère
Morphing Into Odell?", April 12 2006.

103 Simmons, op. cit. The Center for Strategic and
International Studies Workshop, Washington DC, 24
February 2004, saw presentations made by Saudi Aramco
officials and Simmons. These included Mahmoud Abdul
Baqui, "Fifty Year Crude Oil Supply Scenarios: Saudi
Aramco's Perspective", and Nansen Saleri, "Future of
Global Oil Supply".

104 Jim Jarrell, "Another Day in the Desert: A Response to the
Book, 'Twilight in the Desert'", *Geopolitics of Energy*,
October 2005.

105 Michael Lynch, *Crop Circles in the Desert: The Strange
Controversy Over Saudi Oil Production*, International
Research Center for Energy and Economic Development,
Occasional Papers: No. 40, 2005.

106 US Geological Survey World Assessment Team, US
Geological Survey World Petroleum Assessment
2000. This can be found on their website. See here
too Laherrère's critique, "Is USGS 2000 Assessment
Reliable?", op. cit.

107 See Leif Magne Meling, "The Origin of Challenge – Oil
Supply and Demand (Part 1/2)", *MEES*, 12 June 2006. The
author is from Statoil.

108 See "A Reply By C.J. Campbell to Thomas S. Ahlbrandt and J. McCabe, USGS, Published In Geotimes", November 2002 – cited in www.oilcrisis.com.

109 Magne Meling, op. cit.

110 See variously: Michael C. Lynch, *Crying Wolf: Warnings About Oil Supply*, MIT, March 1998; Michael C. Lynch, *The New Pessimism about Petroleum Resources; Debunking the Hubbert Model (and Hubbert Modelers)*, Strategic Energy and Economic Research Inc, 2003; M.A. Adelman and Michael C. Lynch, "Fixed View of Resources Limits Creates Undue Pessimism", *Oil & Gas Journal*, 4 July 1997; Michael C. Lynch, *Crop Circles in the Desert*, op. cit. Also Michael C. Lynch, *The New Geopolitics of Oil*, 2006 (sent to author, 2006).

111 See Esser, op. cit., where these matters are raised, and Peter Stark and Ken Chew, "Global Oil & Gas Resources: The View from the Bottom Up", IHS Energy, 5 May 2004.

112 Wood MacKenzie, "Long-Term Outlook on Oil Market Fundamentals", May 2006.

113 Odell, op. cit.

114 Ibid., p. 142.

115 Erich Zimmermann, *World Resources and Industries*, Harper & Brothers, New York, 1933.

116 For insights on Harry Wassall, including reflections from old friends, see a booklet by Petroconsultants, *Dedicated to Harry Wassall 1921–1995*, produced for the AAPG San Diego Convention in 1996 and put together by Christian Suter, then CEO of the company. Wassall was an

undoubted optimist about the upstream industry and the world.

117 Much on Global Pacific & Partners and its Advisory & Research Practice as well as its management events and Strategy Briefings may be found on www.petro21.com, including conference programmes and past events and speakers under global e-conferences.

118 Leonardo Maugeri, "Two Cheers for Expensive Oil", *Foreign Affairs*, March/April 2006. See also more recently Leonardo Maugeri, *The Age of Oil: The Mythology, History, and Future of the World's Most Controversial Resource*, Praeger, 2006, where a more extensive set of criticisms may be found on Peak Oil in Chapters 16–18.

119 In Part 4 some notions and information have been drawn from extensive reports by Global Pacific & Partners that are available to private clients only by purchase, and hence are not in the public domain. These include the following: "Corporate Oil & 'Global Enemies'"; "National Oil Companies: Strategies, Worldwide Oil Risks & Threats"; "Strategic Intelligence in the 21st Century"; "The Third Scramble for Africa: Origins, Dynamics & Future"; "Global Petroleum Trends". These reports have been produced over the years and are often updated; they are complemented by Annual Strategy Briefings and the firm's Michaelangelo World & Africa Strategy Databank. The firm has undertaken and published similar research over many years on more than 150 national oil companies and over 100 corporate oil entities (super-majors and

independents), as well as on exploration strategies in
over 100 countries in Africa, Asia, Latin America and the
Middle East.

120 Jack Kerfoot, "Exploration & Production Trends:
Implications for Asia", Murphy Oil Malaysia Presentation
at 11th Asia Upstream Conference, Singapore 2005,
Global Pacific & Partners.

121 For some of the problems at PDVSA, see "Oil: We're All
Addicted", *New Statesman*, 17 July 2006.

122 A round-up of the general issues can be found and is well
explained in "Oil's Dark Secret: Special Report National
Oil Companies", *The Economist*, 12 August 2006.

123 See Maugeri, op. cit., p. 220.

124 John D. Grace, *Russian Oil Supply*, Oxford University
Press, 2005.

125 See BP Presentation on CSSB 305, Alaska State
Legislature, Senate Finance Committee, 4 April 2006, by
Angus Walker, commercial vice-president of BP Alaska.

126 Statement by Abdullah S. Jum'ah, president and CEO,
Saudi Aramco, OPEC Annual Meeting, Vienna, 12–13
September 2006.

127 Australian government, "Is the world running out of
oil? A review of the debate", Working Paper 61, Dept of
Transport and Regional Services, Bureau of Transport and
Regional Economics, February 2005.

Notes on further reading

It is neither intended nor necessary to provide any extensive bibliography here. Besides the many sources cited in the notes, reference may be made to several bibliography sources cited by the parties in the debate.

On economic theory, David Warsh is highly recommended for his insights across the discipline and in terms of the history of ideas within economics.

On oil economics, it is best to note the references in publications by Peter R. Odell and also the works of Michael Lynch.

For Peak Oil's case, it is wisest to consult C.J. Campbell's *Oil Crisis*, where an extensive literature is cited, and sources cited in the texts of J.H. Laherrère and in Kenneth S. Deffeyes's *Hubbert's Peak*. A number of general books on oil also contain relevant bibliographic material: see, for instance, Paul Roberts, *The End of Oil*.

An enormous amount of comment on world oil and gas is found on the internet – references have been given in the notes. The official websites of major international energy institutions and agencies provide many key reports and technical documents (see EIA, IEA, OPEC, USGS).

A wealth of geological and other pertinent information on exploration and petroleum by country may be found on the websites of governments, ministries, national oil companies and licensing agencies.

Most public and many private companies from around the world offer important knowledge on the industry and corporate strategies on their websites and in their press releases.

APPENDICES

I

Abbreviations and measures

Abbreviations

AAPG	American Association of Petroleum Geologists
AIM	Alternative Investment Market
ANWR	Alaska National Wildlife Refuge
API	American Petroleum Institute
ASPO	Association for the Study of Peak Oil
BTL	biomass-to-liquids
CBM	coal-bed methane
CEO	chief executive officer
CERA	Cambridge Energy Research Associates
CIA	Central Intelligence Agency
CSIS	Center for Strategic and International Studies
CTL	coal-to-liquids
ECA	Economic Commission for Africa
EEZ	exclusive economic zone
EIA	Energy Information Agency
EOR	enhanced oil recovery
EROEI	energy return on energy invested

EUR	estimated ultimate recovery
GDP	gross domestic product
GTL	gas-to-liquids
HO	heavy oil
IEA	International Energy Agency
IFP	Institut Français du Pétrole
IHS	IHS Energy
ILO	International Labour Organisation
JDZ	Nigeria–São Tomé and Príncipe Joint Development Zone
LNG	liquefied natural gas
NATO	North Atlantic Treaty Organisation
NFW	new field wildcats
NGL	natural gas liquids
NGO	non-governmental organisation
ODAC	Oil Depletion Analysis Centre
OECD	Organisation for Economic Co-operation and Development
OIIP	oil initially in place
OPEC	Organisation of Petroleum Exporting Countries
R/P	reserves/production ratio
SPE	Society of Petroleum Engineers
UNCTAD	United Nations Conference on Trade and Development
UNIDO	United Nations Industrial Development Organisation
URR	ultimate recoverable reserves
USGS	United States Geological Service

WPC	World Petroleum Congress
XHO	extra-heavy oil
YTF	yet-to-find
3-D	three-dimensional seismic

Measures

2-P	proven and probable reserves (50%)
3-P	proven, probable and possible reserves (10%)
BOE	barrel of oil equivalent
BBOE	billion barrels of oil equivalent
MMBLS	million barrels
BBLS	billion barrels
TBLS	trillion barrels
MBPD	thousand barrels per day
MBOPD	thousand barrels of oil per day
MMBOPD	million barrels of oil per day
TCF	trillion cubic feet

2

SPE schema on reserves

			PRODUCTION		
TOTAL PETROLEUM-INITIALLY-IN-PLACE	DISCOVERED PETROLEUM-INITIALLY-IN-PLACE	COMMERCIAL	**RESERVES**		
			Proved	Proved *plus* probable	Proved *plus* probable *plus* possible
		SUB-COMMERCIAL	**CONTINGENT RESOURCES**		
			Low estimate	Best estimate	High estimate
			UNRECOVERABLE		
	UNDISCOVERED PETROLEUM-INITIALLY-IN-PLACE		**PROSPECTIVE RESOURCES**		
			Low estimate	Best estimate	High estimate
			UNRECOVERABLE		

◄ RANGE OF UNCERTAINTY ►

Index

independent licensing for acreage and
    assets 159–60
interventions 188–9
many lack technical and other needed
    capacities 160
many take greater control of the oil
    game 158–9
Nigerian scheme 162
and *OGJ* data 102
oil market projections 32
oil reserve estimates 74
PDVSA in Venezuela 162
Peak Oil's claim that their numbers
    are flawed 215n
policies 112, 131, 135, 137, 141, 142,
    158–62, 195
roadshows and promotions 160
state players hold 90of world reserves
    164
Grace, John D. 178
graphs 115, 136
Great Lakes area, DRC 170
green parties 114
Greenpeace 53
Groppe, Henry 61
Groppe, Long & Little 61
Guinea 173
Guinea-Bissau 171, 175
Gulf of Guinea 113, 175
Gulf of Hormuz 95
Gulf of Mexico 113, 180
Gulf War, first 51

habitat destruction 47
Harper's Magazine 43
heavy oils 41, 69, 90, 159, 171, 181
Heinberg, Richard 31, 54–5
    *The Party's Over: Oil, War and the
    Fate of Industrial Societies* 54
Herold, J.S. 99
Hirsch, Dr Robert 41, 55
HO-XHO (heavy oil-extra heavy oil) 181
Hols, A. 94
Horn of Africa 38
hostage-taking 156
Hotelling, Harold 82
House of Saud 120
House Subcommittee on Energy and Air
    Quality 55

Hubbert, M. King 26, 51, 69, 94, 105–8,
    120, 137
Hubbert's bell curves 26–7, 54
"Hubbert's Peak" 26–7, 105, 107, 108,
    135
human adaptation
    differing views on adaptability 14
    the power of 200
    Stone Age economies 6, 20
human rights 187
hurricanes 156
Hutter, Freddy: *Trendlines* 93
hydrocarbons 181
    calculations based on a per head basis
    36–7
    decline linked to planetary overload
    37

Iceland 34
IFP 59
IHS Energy 136
    data 35, 74, 91, 99, 100–102, 124
imagination 6, 20, 196, 199
Imperial College, London 53
import quotas 34
Incompetence Theory of History 96
India
    "empires of oil" 188
    equity oil 95
    and HO-XHO 181
    new bids in 153
    unconventional oil in 165
    worldwide assets 95
indigenous peoples' rights 187
Indonesia 37, 152, 156, 160, 187
industrial civilisation 46–7
inorganic oil 139
INP 160
Institut Français du Pétrole (IFP) 35
International Association of Energy
    Economists 212n
International Energy Agency (IEA) 29,
    32, 36, 60, 74, 116, 121
international political stability 56
internet 47, 207n
intuitive appreciation 20–21
investment 71, 89, 95, 121, 133, 153,
    161, 184
    Africa 168